Ginseng and Other Medicinal Plants, by

Ginseng and Other Medicinal Plants, by

A. R. (Arthur Robert) Harding This eBook is for the use of anyone anywhere at no cost and with almost no restrictions whatsoever. You may copy it, give it away or re-use it under the terms of the Project Gutenberg License included with this eBook or online at www.gutenberg.org

Title: Ginseng and Other Medicinal Plants A Book of Valuable Information for Growers as Well as Collectors of Medicinal Roots, Barks, Leaves, Etc.

Author: A. R. (Arthur Robert) Harding

Release Date: December 5, 2010 [EBook #34570]

Language: English

Character set encoding: ASCII

*** START OF THIS PROJECT GUTENBERG EBOOK GINSENG AND OTHER MEDICINAL PLANTS ***

Produced by Linda M. Everhart, Blairstown, Missouri (This file was produced from images generously made available by The Internet Archive/American Libraries.)

Ginseng and Other Medicinal Plants

[Frontispiece: Delights in His Ginseng Garden.]

GINSENG AND OTHER MEDICINAL PLANTS

A Book of Valuable Information for Growers as Well as Collectors of Medicinal Roots, Barks, Leaves, Etc.

BY A. R. HARDING

Published by A. R. Harding Publishing Co. Columbus, Ohio

Copyright 1908 By A. R. Harding Pub. Co.

CONTENTS

I. Plants as a Source of Revenue II. List of Plants Having Medicinal Value III. Cultivation of Wild Plants IV. The Story of Ginseng V. Ginseng Habits VI. Cultivation VII. Shading and Blight VIII. Diseases of Ginseng IX. Marketing and Prices X. Letters from Growers XI. General Information XII. Medicinal Qualities XIII. Ginseng in China XIV. Ginseng--Government Description, Etc. XV. Michigan Mint Farm XVI. Miscellaneous Information XVII. Golden Seal Cultivation XVIII. Golden Seal History, Etc. XIX. Growers' Letters XX. Golden Seal--Government Description, Etc. XXI. Cohosh--Black and Blue XXII. Snakeroot--Canada and Virginia XXIII. Pokeweed XXIV. Mayapple XXV. Seneca Snakeroot XXVI. Lady's Slipper XXVII. Forest Roots XXVIII. Forest Plants XXIX. Thicket Plants XXX. Swamp Plants XXXI. Field Plants XXXII. Dry Soil Plants XXXIII. Rich Soil Plants XXXIV.

Medicinal Herbs XXXV. Medicinal Shrubs

LIST OF ILLUSTRATIONS

Delights in His Ginseng Garden Seneca Snake Root (Cultivated) in Blossom Indian Turnip (Wild) Canadian Snake Root (Cultivated) Blood Root (Cultivated) Sarsaparilla Plant (Wild) Ginseng Plants and Roots Garden Grown Ginseng Plants Northern Ginseng Plant in Bloom--June Plan for Ginseng Garden 24 x 40 Feet--Ground Plan One Line, Overhead Dotted A Lath Panel One, Two and Three Year Old Ginseng Roots Ginseng Plants Coming Up Bed of 10,000 Young Ginseng Plants in Forest One Year's Growth of Ginseng Under Lattice Shade A Healthy Looking Ginseng Garden Diseased Ginseng Plants Broken--"Stem Rot" End Root Rot of Seedlings The Beginning of Soft Rot Dug and Dried--Ready for Market A Three Year Old Cultivated Root Bed of Mature Ginseng Plants Under Lattice Some Thrifty Plants--An Ohio Garden New York Grower's Garden Forest Bed of Young "Seng" These Plants However Are Too Thick A Healthy Looking "Garden"--"Yard" Root Resembling Human Body Wild Ginseng Roots Pennsylvania Grower's Garden Ginseng (Panax Quinquefolium) Lady Slipper Young Golden Seal Plant in Bloom Golden Seal Plants Thrifty Golden Seal Plant Golden Seal in an Upland Grove Locust Grove Seal Garden Golden Seal (Hydrastis Canadensis) Flowering Plant and Fruit Golden Seal Rootstock Black Cohosh (Cimicifuga Racemosa), Leaves, Flowering Spikes and Rootstock Blue Cohosh (Caulophyllum Thalictroides) Canada Snakeroot (Asarum Canadense) Virginia Serpentaria (Aristolochia Serpentaria) Pokeweed (Phytolacca Decandra), Flowering and Fruiting Branch Pokeweed Root May-Apple (Podophyllum Pellatum), Upper Portion of Plant with Flower, and Rootstock Seneca Snakeroot (Polygala Senega), Flowering Plant with Root Large Yellow Lady's Slipper (Cypripedium Hirsutum) Bethroot (Trillium Erectum) Culver's Root (Veronica Virginica) Flowering Top and Rootstock Stoneroot (Collinsonia Canadensis) Crawley-Root (Corallorhiza Odontorhiza) Marginal-Fruited Shield-Fern (Dryopteris Marginalis) Goldthread (Coptis Trifolia) Twinleaf (Jeffersonia Diphylla) Plant and Seed Capsule Canada Moonseed (Menispermum Canadense) Wild Turnip (Arisaema Triphyllum) Black Indian Hemp (Apocynum Cannabinum), Flowering Portion, Pods, and Rootstock Chamaelirium (Chamaelirium Luteum) Wild Yam (Dioscorea Villosa)

Skunk-Cabbage (Spathyema Foetida) American Hellebore (Veratrum Viride) Water-Eryngo (Eryngium Yuccifolium) Yellow Jasmine (Gelsensium Sempervirens) Sweet Flag (Acorus Calamus) Blue Flag (Iris Versicolor) Crane's-bill (Geranium Maculatum), Flowering Plant, Showing also Seed Pods and Rootstock Dandelion (Taraxacum Officinale) Soapwort (Saponaria Officinalis) Burdock (Arctium Lappa), Flowering branch and Root Yellow Dock (Rumex Crispus), First Year's Growth Broad-Leaved Dock (Rumex Obtusifolius), Leaf, Fruiting Spike and Root Stillingia (Stillingia Sylvatica), Upper Portion of Plant and Part of Spike Showing Male Plant American Colombo (Frasera Carolinensis), Leaves, Flowers, and Seed Pods Couch-Grass (Agropyron Repens) Echinacea (Brauneria Augustifolia) Aletris (Aletris Farinosa) Wild Indigo (Baptisia Tinctoria), Branch Showing Flowers and Seed Pods Pleurisy Root (Asclepias Tuberosa) Bloodroot (Sanguinaria Canadensis), Flowering Plant with Rootstock Pinkroot (Spigelia Marilandica) Indian Physic (Porteranthus Trifoliatus) Wild Sarsaparilla (Aralia Nudicaulis) American Angelica (Angelica Atropurpurea) Comfrey (Symphytum Officinale) Elecampane (Inula Helenium) Queen-of-the-Meadow (Eupatorium Purpureum) Hydrangea (Hydrangea Arborescens) Oregon Grape (Berberis Aquifolium)

[Illustration: A. R. Harding]

INTRODUCTION

When the price of Ginseng advanced some years ago hundreds engaged in the business who knew little or nothing of farming, plant raising and horticulture. That they largely failed is not to be wondered at. Later many began in a small way and succeeded. Many of these were farmers and gardeners. Others were men who had hunted, trapped and gathered "seng" from boyhood. They therefore knew something of the peculiarities of Ginseng.

It is from the experience of these men that this work is largely made up--writings of those who are in the business.

Golden seal is also attracting considerable attention owing to the rapid increase in price during the early years of the present century. The growing of this plant is given careful attention also.

Many other plants are destined to soon become valuable. A work gotten out by the government--American root drugs--contains a great deal of value in regard habits, range, description, common names, price, uses, etc., etc., so that some of the information contained in this book is taken therefrom. The prices named in the government bulletin which was issued in 1907 were those prevailing at that time--they will vary, in the future, largely according to the supply and demand.

The greatest revenue derived from plants for medicinal purposes is derived from the roots, yet there are certain ones where the leaves and bark are used. Therefore to be complete some space is given to these plants. The digging of the roots, of course, destroys the plant as well as does the peeling of the bark, while leaves secured is clear gain--in other words, if gathered when matured the plant or shrub is not injured and will produce leaves each year.

The amount of root drugs used for medicinal purposes will increase as the medical profession is using of them more and more. Again the number of people in the world is rapidly increasing while the forests (the natural home of root drugs) are becoming less each year. This shows that growers of medicinal roots will find a larger market in the future than in the past.

Those who know something of medicinal plants--"Root Drugs"--can safely embark in their cultivation, for while prices may ease off--go lower--at times, it is reasonably certain that the general trend will be upward as the supply growing wild is rapidly becoming less each year.

A. R. Harding.

CHAPTER I.

PLANTS AS A SOURCE OF REVENUE.

With the single exception of ginseng, the hundred of plants whose roots are used for medical purposes, America is the main market and user. Ginseng is used mainly by the Chinese. The thickly inhabited Chinese Empire is where the American ginseng is principally used. To what uses it is put may be briefly stated, as a superstitious beverage. The roots with certain shapes are carried about the person for charms. The roots resembling the human form being the most valuable.

The most valuable drugs which grow in America are ginseng and golden seal, but there are hundreds of others as well whose leaves, barks, seeds, flowers, etc., have a market value and which could be cultivated or gathered with profit. In this connection an article which appeared in the Hunter-Trader-Trapper, Columbus, Ohio, under the title which heads this chapter is given in full:

To many unacquainted with the nature of the various wild plants which surround them in farm and out-o'-door life, it will be a revelation to learn that the world's supply of crude, botanical (vegetable) drugs are to a large extent gotten from this class of material. There are more than one thousand different kinds in use which are indigenous or naturalized in the United States. Some of these are very valuable and have, since their medicinal properties were discovered, come into use in all parts of the world; others now collected in this country have been brought here and, much like the English sparrow, become in their propagation a nuisance and pest wherever found.

The impression prevails among many that the work of collecting the proper kind, curing and preparing for the market is an occupation to be undertaken only by those having experience and a wide knowledge of their species, uses, etc. It is a fact, though, that everyone, however little he may know of the medicinal value of such things, may easily become familiar enough with this business to successfully collect and prepare for the market many different kinds from the start.

There are very large firms throughout the country whose sole business is for this line of merchandise, and who are at all times anxious to make contracts with parties in the country who will give the work business-like attention, such as would attend the production of other farm articles, and which is so necessary to the success of the work.

If one could visit the buyers of such firms and ask how reliable they have found their sources of supply for the various kinds required, it would provoke much laughter. It is quite true that not more than one in one hundred who write these firms to get an order for some one or more kinds they might supply, ever give it sufficient attention to enable a first shipment to be made. Repeated experiences of this kind have made the average buyer very promptly commit to the nearest waste basket all letters received from those who have not been doing this work in the past, recognizing the utter waste of time in corresponding with those who so far have shown no interest in the work.

The time is ripe for those who are willing to take up this work, seriously giving some time and brains to solving the comparatively easy problems of doing this work at a small cost of time and money and successfully compete for this business, which in many cases is forced to draw supplies from Europe, South America, Africa, and all parts of the world.

From the writer's observation, more of these goods are not collected in this country on account of the false ideas those investigating it have of the amount of money to be made from the work, than from any other reason; they are led to believe that untold wealth lies easily within their reach, requiring only a small effort on their part to obtain it. Many cases may be cited of ones who have laboriously collected, possibly 50 to 100 pounds of an article, and when it was discovered that from one to two dollars per pound was not immediately forthcoming, pronounced the dealer a thief and never again considered the work.

In these days when all crude materials are being bought, manufactured and sold on the closest margins of profit possible, the crude drug business has not escaped, it is therefore only possible to make a reasonable profit in marketing the products of the now useless weeds which confront the farmer as a serious problem at every turn.

CHAPTER I.

To the one putting thought, economy and perseverance in this work, will come profit which is now merely thrown away.

Many herbs, leaves, barks, seeds, roots, berries and flowers are bought in very large quantities, it being the custom of the larger houses to merely place an order with the collector for all he can collect, without restriction. For example, the barks used from the sassafras roots, from the wild cherry tree, white pine tree, elm tree, tansy herb, jimson weed, etc., run into the hundreds of thousand pounds annually, forming very often the basis of many remedies you buy from your druggist.

The idea prevalent with many, who have at any time considered this occupation, that it is necessary to be familiar with the botanical and Latin names of these weeds, must be abolished. When one of the firms referred to receives a letter asking for the price of Rattle Top Root, they at once know that Cimicifuga Racemosa is meant; or if it be Shonny Haw, they readily understand it to mean Viburnum Prunifolium; Jimson Weed as Stramonium Dotura; Indian Tobacco as Lobelia Inflata; Star Roots as Helonias Roots, and so on throughout the entire list of items.

Should an occasion arise when the name by which an article is locally known cannot be understood, a sample sent by mail will soon be the means of making plain to the buyer what is meant.

Among the many items which it is now necessary to import from Germany, Russia, France, Austria and other foreign countries, which might be produced by this country, the more important are: Dandelion Roots, Burdock Roots, Angelica Roots, Asparagus Roots, Red Clover Heads, or blossoms. Corn Silk, Doggrass, Elder Flowers, Horehound Herb, Motherwort Herb, Parsley Root, Parsley Seed, Sage Leaves, Stramonium Leaves or Jamestown Leaves, Yellow Dock Root, together with many others.

Dandelion Roots have at times become so scarce in the markets as to reach a price of 50c per pound as the cost to import it is small there was great profit somewhere.

CHAPTER I.

These items just enumerated would not be worthy of mention were they of small importance. It is true, though, that with one or two exceptions, the amounts annually imported are from one hundred to five hundred thousand pounds or more.

As plentiful as are Red Clover Flowers, this item last fall brought very close to 20c per pound when being purchased in two to ten-ton lots for the Winter's consumption.

For five years past values for all Crude Drugs have advanced in many instances beyond a proportionate advance in the cost of labor, and they bid fair to maintain such a position permanently. It is safe to estimate the average enhancement of values to be at least 100% over this period; those not reaching such an increased price fully made up for by others which have many times doubled in value.

It is beyond the bounds of possibility to pursue in detail all of the facts which might prove interesting regarding this business, but it is important that, to an extent at least, the matter of fluctuations in values be explained before this subject can be ever in a measure complete.

All items embraced in the list of readily marketable items are at times very high in price and other times very low; this is brought about principally by the supply. It is usually the case that an article gradually declines in price, when it has once started, until the price ceases to make its production profitable.

It is then neglected by those formerly gathering it, leaving the natural demand nothing to draw upon except stocks which have accumulated in the hands of dealers. It is more often the case that such stocks are consumed before any one has become aware of the fact that none has been collected for some time, and that nowhere can any be found ready for the market.

Dealers then begin to make inquiry, they urge its collection by those who formerly did it, insisting still upon paying only the old price. The situation becomes acute; the small lots held are not released until a fabulous price may be realized, thus establishing a very much higher market. Very soon the advanced prices reach the collector, offers are rapidly made him at higher and higher prices, until finally every one in the district is attracted by the high and profitable figures being offered. It is right here that every careful person concerned needs to be doubly careful else, in the inevitable drop in prices caused by the over-production which as a matter of course follows, he will lose money. It will probably take two to five years then for this operation to repeat itself with these items, which have after this declined even to lower figures than before.

In the meantime attention is directed to others undergoing the same experience. A thorough understanding of these circumstances and proper heed given to them, will save much for the collector and make him win in the majority of cases.

Books and other information can be had by writing to the manufacturers and dealers whose advertisements may be found in this and other papers.

CHAPTER II.

LIST OF PLANTS HAVING MEDICINAL VALUE.

The list of American Weeds and Plants as published under above heading having medicinal value and the parts used will be of especial value to the beginner, whether as a grower, collector or dealer.

The supply and demand of medicinal plants changes, but the following have been in constant demand for years. The name or names in parenthesis are also applied to the root, bark, berry, plant, vines, etc., as mentioned:

Balm Gilead (Balsam Poplar)--The Buds. Bayberry (Wax-Myrtle)--The Bark of Root. Black Cohosh (Black Snake Root)--The Root with Rootlets. Black Haw (Viburnum. Sloe.)--The Bark of Root. The Bark of Tree. Black Indian Hemp (Canadian Hemp)--The Root. Blood Root--The Root with Fibre. The Root with no Fibre. Blue Cohosh (Papoose Root. Squaw Root)--The Root. Blue Flag (Larger Blue Flag)--The Root. Burdock--The Root. The Seed. Cascara Sagrada (Chittem Bark)--Bark of Tree. Clover, Red--The Blossoms. Corn Silk Cotton Root--The Bark of Root. Cramp Root (Cranberry Tree. High Bush Cranberry)--The Bark of Tree. Culver's Root (Black Root)--The Root. Dandelion--The Root. Deer Tongue--The Leaves. Elder--The Dried Ripe Berries. The Flowers. Elecampane--The Root, cut into slices. Elm (Slippery Elm)--The Bark, deprived of the brown, outside layer. Fringe Tree--The Bark of Root. Gelsemium (Yellow Jasmine) (Carolina Jasmine)--The Root. Ginseng--The Root. Golden Seal (Yellow Root. Yellow Puccoon. Orange Root. Indian Dye. Indian Turmeric)--The Root. Gold Thread (Three-leaved Gold Thread)--The Herb. Hops--These should be collected and packed in such a manner as to retain all of the yellow powder (lupulin.) Hydrangea--The Root. Indian Hemp, Black (See Black Indian Hemp) Lady Slipper (Moccasin-Flower. Large Yellow Lady Slipper. American Valerian)--The Root, with Rootlets. Lobelia (Indian Tobacco)--The Herb. The Seed. Mandrake (May-apple)--The Root. Nettle--The Herb. Passion Flower--The Herb. Pipsissewa (Prince's Pine)--The Vine. Poke--The Berries. The Root. Prickly Ash (Toothache Tree. Angelica Tree. Suterberry. Pepper Wood. Tea Ash)--The Bark. The Berry.

CHAPTER II.

Sassafras--The Bark of the Root. The Pith. Saw Palmetto--The Berries. Scullcap--The Herb.

[Illustration: Senega Snake Root (Cultivated) in Blossom.]

Snake Root, Virginia (Birthwort-Serpentaria)--The Root. Snake Root, Canada (Asarabacca. Wild Ginger. So-called Coltfoot Root)--The Root. Spruce Gum--Clean Gum only. Squaw Vine (Partridge Berry)--The Herb. Star Root (See Unicorn False) Star Grass (See Unicorn True) Stillingia (Queen's Delight)--The Root. Stramonium (Jamestown-weed. Jimson-weed. Thorn-apple)--The Leaves. The Seed. Unicorn True (Star Grass. Blazing Star. Mealy Starwort. Colic Root)--The Root. Unicorn False (Star Root. Starwort)--The Root. Wahoo (Strawberry Tree. Indian Arrow. Burning Bush. Spindle Tree. Pegwood. Bitter Ash)--The Bark of Root. The Bark of Tree. White Pine (Deal Pine. Soft Deal Pine)--The Bark of Tree, Rossed. Wild Cherry--The thin Green Bark, and thick Bark Rossed. The dried Cherries. Wild Indigo (Horsefly Weed. Rattle-bush. Indigo Weed. Yellow Indigo. Clover Broom)--The Root. Wormseed, American (Stinking Weed. Jesuit Tea. Jerusalem Tea. Jerusalem Oak)--The Seed. Wild Yam (Colic Root. China Root. Devil's Bones)--The Root. Yellow Dock (Sour Dock. Narrow Dock. Curled Dock)--The Root.

The following are used in limited quantities only:

Arbor Vitae (White Cedar)--The Leafy Tips. Balmony (Turtle-head. Snakehead)--The Herb, free from large Stalks. Beth Root (Trillium Erectum. Wake Robin. Birth-root)--The Root. Birch Bark (Cherry Birch. Sweet Birch. Black Birch. Black Root (see culvers root)--The Bark of Tree. Blackberry (High Blackberry)--The Bark of Root. Black Willow--The Bark. The Buds. Boneset (Thoroughwort)--The Herb, free from large Stems. Broom Corn--The Seed. Broom Top (Scotch Broom)--The Flowering Tops. Bugle Weed (Water Horehound) The Herb, free from large Stems. Butternut--Bark of Root. Catnip--The Herb. Chestnut--The Leaves, collected in September or October while still green. Chicory (Succory)--The Root, cut into slices (Cross section.) Corn Ergot (Corn Smut)--The Fungus, replacing the grains of corn. False Bittersweet (Shrubby Bittersweet. Climbing Bittersweet. Wax-wort. Staff-tree)--The Bark of Tree. Garden Lettuce--The Leaves.

CHAPTER II.

Geranium (Cranesbill)--The Root of the wild Herb. Gravel Plant (May Flower. Ground Laurel. Trailing Arbutus)--The Leaves. Great Celandine (Garden Celandine)--Entire plant. Hellebore, False (Adonis Vernalis)--The Root. Hemlock--The Bark. The Gum. Horse Nettle--The Berries. The Root. Huckleberry--The Dried Berry. Life Everlasting (Common Everlasting. Cudweed)--The Herb. Life Root Plant (Rag-wort)--The Herb. Lovage--The Root. Maiden Hair--The Fern. Milkweed (Pleurisy Root)--The Root cut into Sections lengthwise. Motherwort--The Herb. Mountain Ash (Mountain Laurel (See Sheep Laurel)--The Bark of Tree. Mullein (Common Mullein)--The Leaves. Pennyroyal--The Herb. Peppermint The Leaves.--The Herb. Pitcher Plant (Side-Saddle Plant. Fly Trap. Huntsman Cup. Water Cup)--The Plant. Plantain (Rib-grass. Rib-wort. Ripple-grass)--The Leaves. Poison Oak (Poison Ivy)--The Leaves. Pumpkin--The Seed. Queen of the Meadow (Joe-Pye-Weed. Trumpet-Weed)--The Root. Ragweed (Wild Red Raspberry)--The Leaves. Rosinweed (Polar plant. Compass plant)--The Root. Rue--The Herb. Sage--The Leaves. Scouring Rush (Horsetail)--The Herb. Sheep Laurel (Laurel. Mountain Laurel. Broad-leafed Laurel. Calico Bush. Spoon Wood)--The Leaves. Sheep Sorrel (Field Sorrel)--The Leaves. Shepherd's Purse--The Herb. Skunk Cabbage--The Root. Spikenard--The Root. Stone Root--The Root. Tag Alder--The Bark. Tansy (Trailing Arbutus. See Gravel Plant)--The Herb. Veratrum Viride (Green Hellebore. American Hellebore)--The Root. Vervain (Blue Vervain)--The Herb. Virginia Stone Crop (Dutch Stone Crop) Wafer Ash (Hop Tree. Swamp Dogwood. Stinking Ash. Scrubby Trefoil. Ague Bark)--The Bark of Root. Water Avens (Throat Root. Cure All. Evan's Root. Indian Chocolate. Chocolate Root. Bennett Root)--The Root. Water Eryngo (Button Snake Root. Corn Snake Root. Rattle Snake's Weed)--The Root. Water Hemlock (Spotted Parsley. Spotted Hemlock. Poison Parsley. Poison Hemlock. Poison Snake Weed. Beaver Poison)--The Herb. Watermelon--The Seed. Water Pepper (Smart Weed. Arsmart)--The Herb. Water Ash--The Bark of Tree. White Oak (Tanners Bark)--The Bark of Tree, Rossed. White Ash--The Bark of Tree. White Poplar (Trembling Poplar. Aspen. Quaking Asp)--The Bark of Tree. Wild Lettuce (Wild Opium Lettuce. Snake Weed. Trumpet Weed)--The Leaves.

[Illustration: Indian Turnip (Wild).]

CHAPTER II.

Wild Turnip (Indian Turnip. Jack-in-the-Pulpit. Pepper Turnip. Swamp Turnip)--The Root, sliced. Wintergreen (Checkerberry. Partridge Berry. Teaberry. Deerberry)--The Leaves. Witch Hazel (Striped Alder. Spotted Alder. Hazelnut)--The Bark. The Leaves. Yarrow (Milfoil. Thousand Leaf)--The Herb. Yellow Parilla (Moon Seed. Texas Sarsaparilla)--The Root. Yerba Santa (Mountain Balm. Gum Plant. Tar Weed)--The Leaves.

CHAPTER III.

CULTIVATION OF WILD PLANTS.

The leading botanical roots in demand by the drug trade are the following, to-wit: Ginseng, Golden Seal, Senega or Seneca Snake Root, Serpentaria or Virginia Snake Root, Wild Ginger or Canada Snake Root, Mandrake or Mayapple, Pink Root, Blood Root, Lady Slipper, Black Root, Poke Root and the Docks. Most of these are found in abundance in their natural habitat, and the prices paid for the crude drugs will not, as yet, tempt many persons to gather the roots, wash, cure, and market them, much less attempt their culture. But Ginseng, Golden Seal, Senega, Serpentaria and Wild Ginger are becoming very scarce, and the prices paid for these roots will induce persons interested in them to study their several natures, manner of growth, natural habitat, methods of propagation, cultivation, etc.

This opens up a new field of industry to persons having the natural aptitude for such work. Of course, the soil and environment must be congenial to the plant grown. A field that would raise an abundance of corn, cotton, or wheat would not raise Ginseng or Golden Seal at all. Yet these plants grown as their natures demand, and by one who "knows," will yield a thousand times more value per acre than corn, cotton or wheat. A very small Ginseng garden is worth quite an acreage of wheat. I have not as yet marketed any cultivated Ginseng. It is too precious and of too much value as a yielder of seeds to dig for the market.

Some years ago I dug and marketed, writes a West Virginia party, the Golden Seal growing in a small plot, ten feet wide by thirty feet long, as a test, to see if the cultivation of this plant would pay. I found that it paid extremely well, although I made this test at a great loss. This bed had been set three years. In setting I used about three times as much ground as was needed, as the plants were set in rows eighteen inches apart and about one foot apart in the rows. The rows should have been one foot apart, and the plants about six inches apart in the rows, or less. I dug the plants in the fall about the time the tops were drying down, washed them clean, dried them carefully in the shade and sold them to a man in the city of Huntington, W Va. He paid me $1.00 per

CHAPTER III.   16

pound and the patch brought me $11.60, or at the rate of $1,684.32 per acre, by actual measure and test.

[Illustration: Canadian Snake Root (Cultivated).]

This experiment opened my eyes very wide. The patch had cost me practically nothing, and taking this view only, had paid "extremely well." But, I said, "I made this test at a great loss," which is true, taking the proper view of the case. Suppose I had cut those roots up into pieces for propagation, and stratified them in boxes of sandy loam through the winter, and when the buds formed on them carefully set them in well prepared beds. I would now have a little growing gold mine. The price has been $1.75 for such stock, or 75% more than when I sold, making an acre of such stuff worth $2,948.56. The $11.60 worth of stock would have set an acre, or nearly so. So my experiment was a great loss, taking this view of it.

I am raising, in a small way, Ginseng, Lady Slipper, Wild Ginger and Virginia Snake Root, and am having very good success with all of it. I am also experimenting with some flowering plants, such as Sweet Harbinger, Hepatica, Blood Root, and Blue Bell. I am trying to propagate and grow some shrubs and trees to be used as yard and cemetery trees. Of these my most interesting one is the American Christmas Holly. I have not made much headway with it yet, but I am not discouraged. I know more about it than when I began, and think I shall succeed. There is good demand for Holly at Christmas time, and I can find ready sale for all I can get. I think the plants should sell well, as it makes a beautiful shrub. I think the time has come when the Ginseng and Golden Seal of commerce and medicine will practically all come from the gardens of the cultivators of these plants. I do not see any danger of overproduction. The demand is great and is increasing year by year. Of course, like the rising of a river, the price may ebb and flow, somewhat, but it is constantly going up.

[Illustration: Blood Root (Cultivated).]

The information contained in the following pages about the habits, range, description and price of scores of root drugs will help hundreds to distinguish the valuable plants from the worthless. In most instances

CHAPTER III.

a good photo of the plant and root is given. As Ginseng and Golden Seal are the most valuable, instructions for the cultivation and marketing of same is given in detail. Any root can be successfully grown if the would-be grower will only give close attention to the kind of soil, shade, etc., under which the plant flourishes in its native state.

[Illustration: Sarsaparilla Plant (Wild).]

Detailed methods of growing Ginseng and Golden Seal are given from which it will be learned that the most successful ones are those who are cultivating these plants under conditions as near those as possible which the plants enjoy when growing wild in the forests. Note carefully the nature of the soil, how much sunlight gets to the plants, how much leaf mould and other mulch at the various seasons of the year.

It has been proven that Ginseng and Golden Seal do best when cultivated as near to nature as possible. It is therefore reasonable to assume that all other roots which grow wild and have a cash value, for medicinal and other purposes, will do best when "cultivated" or handled as near as possible under conditions which they thrived when wild in the forests.

Many "root drugs" which at this time are not very valuable--bringing only a few cents a pound--will advance in price and those who wish to engage in the medicinal root growing business can do so with reasonable assurance that prices will advance for the supply growing wild is dwindling smaller and smaller each year. Look at the prices paid for Ginseng and Golden Seal in 1908 and compare with ten years prior or 1898. Who knows but that in the near future an advance of hundreds of per cent. will have been scored on wild turnip, lady's slipper, crawley root, Canada snakeroot, serpentaria (known also as Virginia and Texas snakeroot), yellow dock, black cohosh, Oregon grape, blue cohosh, twinleaf, mayapple, Canada moonseed, blood-root, hydrangea, crane's bill, seneca snakeroot, wild sarsaparilla, pinkroot, black Indian hemp, pleurisy-root, culvers root, dandelion, etc., etc.?

Of course it will be best to grow only the more valuable roots, but at the same time a small patch of one or more of those mentioned above

CHAPTER III.

may prove a profitable investment. None of these are apt to command the high price of Ginseng, but the grower must remember that it takes Ginseng some years to produce roots of marketable size, while many other plants produce marketable roots in a year.

There are thousands of land owners in all parts of America that can make money by gathering the roots, plants and barks now growing on their premises. If care is taken to only dig and collect the best specimens an income for years can be had.

CHAPTER IV.

THE STORY OF GINSENG.

History and science have their romances as vivid and as fascinating as any in the realms of fiction. No story ever told has surpassed in interest the history of this mysterious plant Ginseng; the root that for nearly 200 years has been an important article of export to China.

Until a few years ago not one in a hundred intelligent Americans living in cities and towns, ever heard of the plant, and those in the wilder parts of the country who dug and sold the roots could tell nothing of its history and use. Their forefathers had dug and sold Ginseng. They merely followed the old custom.

The natural range of Ginseng growing wild in the United States is north to the Canadian line, embracing all the states of Maine, New Hampshire, Vermont, Massachusetts, Connecticut, Rhode Island, Delaware, New York, Pennsylvania, New Jersey, Maryland, Ohio, West Virginia, Virginia, Indiana, Illinois, Michigan, Wisconsin, Kentucky and Tennessee. It is also found in a greater part of the following states: Minnesota, Iowa, Missouri, North and South Carolina, Georgia and Alabama. Until recently the plant was found growing wild in the above states in abundance, especially those states touched by the Allegheny mountains. The plant is also found in Ontario and Quebec, Canada, but has become scarce there also, owing to persistent hunting. It also grows sparingly in the states west of and bordering on the Mississippi river.

Ginseng in the United States was not considered of any medical value until about 1905, but in China it is and has been highly prized for medical purposes and large quantities of the root are exported to that country. It is indeed doubtful if the root has much if any medical value, and the fact that the Chinese prefer roots that resemble, somewhat, the human body, only goes to prove that their use of the root is rather from superstition than real value.

Of late years Ginseng is being cultivated by the Chinese in that country, but the root does not attain the size that it does in America,

CHAPTER IV.

and the plant from this side will, no doubt, continue to be exported in large quantities.

New York and San Francisco are the two leading cities from which exports are made to China, and in each of these places are many large dealers who annually collect hundreds of thousands of dollars' worth. The most valuable Ginseng grows in New York, the New England states and northern Pennsylvania. The root from southern sections sells at from fifty cents to one dollar per pound less.

Ginseng in the wild or natural state grows largely in beech, sugar and poplar forests and prefers a damp soil. The appearance of Ginseng when young resembles somewhat newly sprouted beans; the plant only grows a few inches the first year. In the fall the stem dies and in the spring the stalk grows up again. The height of the full grown stalk is from eighteen to twenty inches, altho they sometimes grow higher. The berries and seed are crimson (scarlet) color when ripe in the fall. For three or four years the wild plants are small, and unless one has a practical eye will escape notice, but professional diggers have so persistently scoured the hills that in sections where a few years ago it was abundant, it is now extinct.

While the palmy days of digging were on, it was a novel occupation and the "seng diggers," as they are commonly called, go into the woods armed with a small mattock and sack, and the search for the valuable plant begins. Ginseng usually grows in patches and these spots are well known to the mountain residents. Often scores of pounds of root are taken from one patch, and the occupation is a very profitable one. The women as well as the men hunt Ginseng, and the stalk is well known to all mountain lads and lassies. Ginseng grows in a rich, black soil, and is more commonly found on the hillsides than in the lowlands.

[Illustration: Ginseng Plant and Roots.]

Few are the mountain residents who do not devote some of their time to hunting this valuable plant, and in the mountain farm houses there are now many hundred pounds of the article laid away waiting the market. While the fall is the favorite time for Ginseng hunting, it is

CHAPTER IV.   21

carried on all summer. When a patch of the root is found the hunter loses no time in digging it. To leave it until the fall would be to lose it, for undoubtedly some other hunter would find the patch and dig it.

How this odd commerce with China arose is in itself remarkable. Many, many years ago a Catholic priest, one who had long served in China, came as a missionary to the wilds of Canada. Here in the forest he noted a plant bearing close resemblance to one much valued as a medicine by the Chinese. A few roots were gathered and sent as a sample to China, and many months afterwards the ships brought back the welcome news that the Chinamen would buy the roots.

Early in its history the value of Ginseng as a cultivated crop was recognized, and repeated efforts made for its propagation. Each attempt ended in failure. It became an accepted fact with the people that Ginseng could not be grown. Now these experimenters were not botanists, and consequently they failed to note some very simple yet essential requirements of the plant. About 1890 experiments were renewed. This time by skilled and competent men who quickly learned that the plant would thrive only under its native forest conditions, ample shade, and a loose, mellow soil, rich in humus, or decayed vegetable matter. As has since been shown by the success of the growers. Ginseng is easily grown, and responds readily to proper care and attention. Under right conditions the cultivated roots are much larger and finer, and grow more quickly than the wild ones.

It may be stated in passing, that Chinese Ginseng is not quite the same thing as that found in America, but is a variety called Panax Ginseng, while ours is Panax Quinquefolia. The chemists say, however, that so far as analysis shows, both have practically the same properties. It was originally distributed over a wide area, being found everywhere in the eastern part of the United States and Canada where soil and locality were favorable.

Ginseng has an annual stalk and perennial root. The first year the foliage does not closely resemble the mature plant, having only three leaves. It is usually in its third year that it assumes the characteristic leaves of maturity and becomes a seed-bearer. The photos which accompany give a more accurate idea of the plant's appearance than

CHAPTER IV.

is possible from a written description. The plants bloom very quickly after sprouting and the berries mature in August and September in most localities. When ripe, the berries are a rich deep crimson and contain usually two seeds each.

The seeds are peculiar in that it usually takes them about eighteen months to germinate and if allowed to become dry in the meantime, the vitality will be destroyed.

Western authorities have heretofore placed little value on Ginseng as a curative agent, but a number of recent investigations seem to reverse this opinion. The Chinese, however, have always placed the highest value upon it and millions have used and esteemed it for untold centuries. Its preparation and uses have never been fully understood by western people.

Our Consuls in China have at various times furnished our government with very full reports of its high value and universal use in the "Flowery Kingdom." From these we learn that "Imperial Ginseng," the highest grade grown in the royal parks and gardens, is jealously watched and is worth from $40.00 to $200.00 per pound. Of course its use is limited to the upper circle of China's four hundred. The next quality comes from Korea and is valued at $15.00 to $35.00 per pound. Its use is also limited to the lucky few. The third grade includes American Ginseng and is the great staple kind. It is used by every one of China's swarming millions who can possibly raise the price. The fourth grade is Japanese Ginseng and is used by those who can do no better.

Mr. Wildman, of Hong Kong, says: "The market for a good article is practically unlimited. There are four hundred million Chinese and all to some extent use Ginseng. If they can once become satisfied with the results obtained from the tea made from American Ginseng, the yearly demand will run up into the millions of dollars worth." Another curious fact is that the Chinese highly prize certain peculiar shapes among these roots especially those resembling the human form. For such they gladly pay fabulous prices, sometimes six hundred times its weight in silver. The rare shapes are not used as medicine but kept as a charm, very much as some Americans keep a rabbit's foot for luck.

CHAPTER IV.

Sir Edwin Arnold, that famous writer and student of Eastern peoples, says of its medicinal values: "According to the Chinaman, Ginseng is the best and most potent of cordials, of stimulants, of tonics, of stomachics, cardiacs, febrifuges, and, above all, will best renovate and reinvigorate failing forces. It fills the heart with hilarity while its occasional use will, it is said, add a decade of years to the ordinary human life. Can all these millions of Orientals, all those many generations of men, who have boiled Ginseng in silver kettles and have praised heaven for its many benefits, have been totally deceived? Was the world ever quite mistaken when half of it believed in something never puffed, nowhere advertised and not yet fallen to the fate of a Trust, a Combine or a Corner?"

It has been asked why the Chinese do not grow their own Ginseng. In reply it may be said that America supplies but a very small part indeed of the Ginseng used in China. The bulk comes from Korea and Manchuria, two provinces belonging to China, or at least which did belong to her until the recent Eastern troubles.

Again, Ginseng requires practically a virgin soil, and as China proper has been the home of teeming millions for thousands of years, one readily sees that necessary conditions for the plant hardly exist in that old and crowded country.

CHAPTER V.

GINSENG HABITS.

A few years ago Ginseng could be found in nearly every woods and thicket in the country. Today conditions are quite different. Ginseng has become a scarce article. The decrease in the annual crop of the wild root will undoubtedly be very rapid from this on. The continued search for the root in every nook and corner in the country, coupled with the decrease in the forest and thicket area of the country, must in a few years exterminate the wild root entirely.

To what extent the cultivated article in the meantime can supplant the decrease in the production of the wild root, is yet to be demonstrated. The most important points in domesticating the root, to my opinion, is providing shade, a necessary condition for the growth of Ginseng, and to find a fertilizer suitable for the root to produce a rapid growth. If these two conditions can be complied with, proper shade and proper fertilizing, the cultivation of the root is simplified. Now the larger wild roots are found in clay soil and not in rich loam. It seems reasonably certain that the suitable elements for the growth of the root is found in clay soil.

The "seng" digger often finds many roots close to the growing stalk, which had not sent up a shoot that year. For how many years the root may lie dormant is not known, nor is it known whether this is caused by lack of cultivation. I have noticed that the cultivated plant did not fail to sprout for five consecutive years. Whether it will fail the sixth year or the tenth is yet unknown. The seed of Ginseng does not sprout or germinate until the second year, when a slender stalk with two or three leaves puts in an appearance. Then as the stalk increases in size from year to year, it finally becomes quite a sizable shrub of one main stalk, from which branch three, four, or even more prongs; the three and four prongs being more common. A stalk of "seng" with eight well arranged prongs, four of which were vertically placed over four others, was found in this section (Southern Ohio) some years ago. This was quite an oddity in the general arrangement of the plant.

CHAPTER V.

Ginseng is a plant found growing wild in the deep shaded forests and on the hillsides thruout the United States and Canada. Less than a score of years ago Ginseng was looked upon as a plant that could not be cultivated, but today we find it is successfully grown in many states. It is surprising what rapid improvements have been made in this valuable root under cultivation. The average cultivated root now of three or four years of age, will outweigh the average wild root of thirty or forty years.

When my brother and I embarked in the enterprise, writes one of the pioneers in the business, of raising Ginseng, we thought it would take twenty years to mature a crop instead of three or four as we are doing today. At that time we knew of no other person growing it and from then until the present time we have continually experimented, turning failures to success. We have worked from darkness to light, so to speak.

In the forests of Central New York, the plant is most abundant on hillsides sloping north and east, and in limestone soils where basswood or butternut predominate. Like all root crops, Ginseng delights in a light, loose soil, with a porous subsoil.

If a cultivated plant from some of our oldest grown seed and a wild one be set side by side without shading, the cultivated one will stand three times as long as the wild one before succumbing to excessive sunlight. If a germinated seed from a cultivated plant were placed side by side under our best mode of cultivation, the plant of the cultivated seed at the end of five years, would not only be heavier in the root but would also produce more seed.

In choosing a location for a Ginseng garden, remember the most favorable conditions for the plant or seed bed are a rich loamy soil, as you will notice in the home of the wild plant. You will not find it on low, wet ground or where the Water stands any length of time, it won't grow with wet feet; it wants well drained soil. A first-class location is on land that slopes to the east or north, and on ground that is level and good. Other slopes are all right, but not as good as the first mentioned. It does not do so well under trees, as the roots and fibers from them draw the moisture from the plant and retard its growth.

CHAPTER V.

[Illustration: Garden Grown Ginseng Plant.]

The variety of soil is so much different in the United States that it is a hard matter to give instructions that would be correct for all places. The best is land of a sandy loam, as I have mentioned before. Clay land can be used and will make good gardens by mixing leaf mold, rotten wood and leaves and some lighter soil, pulverize and work it thru thoroughly. Pick out all sticks and stones that would interfere with the plants.

Ginseng is a most peculiar plant. It has held a place of high esteem among the Chinese from time immemorial. It hides away from man with seeming intelligence. It is shy of cultivation, the seed germinating in eighteen months as a rule, from the time of ripening and planting. If the seeds become dry they lose, to a certain extent, their germinating power.

The young plant is very weak and of remarkably slow growth. It thrives only in virgin soil, and is very choice in its selection of a place to grow. Remove the soil to another place or cultivate it in any way and it loses its charm for producing this most fastidious plant.

It has a record upon which it keeps its age, or years of its growth, for it passes a great many years in the ground, dormant. I have counted the age upon the record stem of small roots and found their age to be from 30 to 60 years. No plant with which I am acquainted grows as slowly as Ginseng.

A great many superstitious notions are held by the people, generally, in regard to Ginseng. I think it is these natural peculiarities of the plant, together with the fancied resemblance of the root to man, and, also probably its aromatic odor that gives it its charm and value. Destroy it from the earth and the Materia Medica of civilization would lose nothing.

I notice that the cultivated root is not so high in price by some two dollars as the wild root. If the root is grown in natural environment and by natural cultivation, i. e., just let it grow, no Chinaman can tell it from the wild root.

CHAPTER V.

We have at present, writes a grower, in our Ginseng patch about 3,500 plants and will this year get quite a lot of excellent seed. Our Ginseng garden is on a flat or bench on a north hillside near the top, that was never cleared. The soil is a sandy loam and in exposure and quality naturally adapted to the growth of this plant. The natural growth of timber is walnut, both black and white, oak, red bud, dogwood, sugar, maple, lin, poplar and some other varieties.

We cultivate by letting the leaves from the trees drop down upon the bed in the fall as a mulch and then in the early spring we burn the leaves off the bed. Our plants seem to like this treatment very well. They are of that good Ginseng color which all Ginseng diggers recognize as indicative of good sized, healthy roots.

[Illustration: Northern Ginseng Plant in Bloom--June.]

I have had much experience in hunting the wild Ginseng roots, says another, and have been a close observer of its habits, conditions, etc. High shade is best with about one-half sun. The root is found mostly where there is good ventilation and drainage. A sandy porous loam produces best roots. Plants in dense shade fail to produce seed in proportion to the density of the shade. In high one-half shade they produce heavy crops of seed. Coarse leaves that hold water will cause disease in rainy seasons. No leaves or mulch make stalks too low and stunted.

Ginseng is very wise and knows its own age. This age the plant shows in two ways. First, by the style of the foliage which changes each year until it is four years old. Second, the age can be determined by counting the scars on the neck of the bud-stem. Each year the stalk which carries the leaves and berries, goes down, leaving a scar on the neck or perennial root from which it grew. A new bud forms opposite and a little above the old one each year. Counting these stalk scars will give the age of the plant.

I have seen some very old roots and have been told that roots with fifty scars have been dug. The leaf on a seedling is formed of three small parts on a stem, growing directly out of a perennial root and during the first year it remains that shape. The second year the stem forks at the top and each fork bears two leaves, each being formed of five parts. The third year the stem forks three ways and bears three leaves, each formed of five parts, much like the Virginia creeper.

Now the plant begins to show signs of bearing seed and a small button-shaped cluster of green berries can be seen growing in the forks of the stalk at the base of the leaf stem. The fourth year the perennial stalk grows as large around as an ordinary lead pencil and from one foot to twenty inches high. It branches four ways, and has four beautiful five-pointed leaves, with a large well-formed cluster of berries in the center. After the middle of June a pale green blossom forms on the top of each berry. The berries grow as large as a cherry pit and contain two or three flat hard seeds. In September they turn a beautiful red and are very attractive to birds and squirrels. They may be gathered each day as they ripen and should be planted directly in a bed, or put in a box of damp, clean sand and safely stored. If put directly in the ground they will sprout the first year, which advantage would be lost if stored dry.

A word to trappers about wild roots. When you find a plant gather the seed, and unless you want to plant them in your garden, bury them in the berry about an inch or inch and a half deep in some good, rich, shady place, one berry in each spot. Thus you will have plants to dig in later years, you and those who come after you. Look for it in the autumn after it has had time to mature its berries. Do not take up the little plants which have not yet become seed bearers.

CHAPTER VI.

CULTIVATION.

The forest is the home of the Ginseng plant and the closer we follow nature the better results we get. I am growing it now under artificial shade; also in the forest with natural shade, says an Ohio party. A good shade is made by setting posts in the ground, nail cross-pieces on these, then cover with brush. You must keep out the sun and let in the rain and this will do both. Another good shade is made by nailing laths across, allowing them to be one-half inch apart. This will allow the rain to pass thru and will keep the sun out. Always when using lath for shade allow them to run east and west, then the sun can't shine between them.

In selecting ground for location of a Ginseng garden, the north side of a hill is best, altho where the ground is level it will grow well. Don't select a low marshy piece of ground nor a piece too high, all you want is ground with a good drainage and moisture. It is the opinion of some people that in a few years the market will be glutted by those growing it for sale. I will venture to say that I don't think we can grow enough in fifty years to over-run the market. The demand is so great and the supply so scarce it will be a long time before the market will be affected by the cultivated root.

The market has been kept up entirely in the past by the wild root, but it has been so carelessly gathered that it is almost entirely exhausted, so in order to supply this demand we must cultivate this crop. I prepare my beds five feet wide and as long as convenient. I commence by covering ground with a layer of good, rich, loose dirt from the woods or well-rotted manure. Then I spade it up, turning under the rich dirt. Then I cover with another layer of the same kind of dirt in which I plant my seed and roots.

After I have them planted I cover the beds over with a layer of leaves or straw to hold the moisture, which I leave on all winter to protect them from the cold. In the spring I remove a part of the leaves (not all), they will come up thru the leaves as they do in their wild forest.

CHAPTER VI.

All the attention Ginseng needs after planting is to keep the weeds out of the beds. Never work the soil after planting or you will disturb the roots. It is a wild plant and we must follow nature as near as possible.

Ginseng can be profitably grown on small plots if it is cared for properly. There are three things influencing its growth. They are soil, shade and treatment. In its wild state the plant is found growing in rich leaf mold of a shady wood. So in cultivation one must conform to many of the same conditions in which the plant is found growing wild.

In starting a bed of Ginseng the first thing to be considered is the selection of soil. Tho your soil be very rich it is a good plan to cover it with three or four inches of leaf mold and spade about ten inches deep so that the two soils will be well mixed. Artificial shade is preferable at all times because trees take nearly all the moisture and strength out of the soil.

When the bed is well fitted, seed may be sown or plants may be set out. The latter is the quicker way to obtain results. If seeds are sown the young grower is apt to become discouraged before he sees any signs of growth, as it requires eighteen months for their germination. The cheapest way to get plants is to learn to recognize them at sight, then go to the woods and try to find them. With a little practice you will be able to tell them at some distance. Much care should be taken in removing the plant from the soil. The fewer fibers you break from the root, the more likely it will be to grow. Care should also be taken not to break the bud on top of the root. It is the stalk of the plant starting for the next year, and is very noticeable after June 1st. If it be broken or harmed the root will have no stalk the next season.

It is best to start a Ginseng garden on a well drained piece of land, says a Dodge County, Wisconsin, grower. Run the beds the way the hill slopes. Beds should only be four to five feet wide so that they can be reached, for walking on the beds is objectionable. Make your walks about from four to six inches below the beds, for an undrained bed will produce "root rot." The ground should be very rich and "mulchy." Use well rotted horse manure in preparing the beds, for fresh manure will heat and hurt the plants. Use plenty of woods dirt, but very little manure of any kind.

CHAPTER VI.

Set plants about six inches each way, and if you want to increase the size of the root, pinch off the seed bulb. In the fall when the tops have died down, cover the beds about two inches deep with dead leaves from the woods. We make our shades out of one-inch strips three inches wide and common lath. The north and west fence should be more tight to keep cold winds out. Eastern and southern side tight, two feet from the ground. From the two feet to top you may use ordinary staves from salt barrels or so nailed one inch apart. Have your Ginseng garden close to the house, for Ginseng thieves become numerous.

I was raised in the country on a farm and as near to nature as it is possible to get, and have known a great deal of Ginseng from my youth up. Twenty-five years ago it was 75 cents a pound, and now it is worth ten times as much. Every one with any experience in such matters knows that if radishes or turnips are planted in rich, old soil that has been highly fertilized they will grow large and will be strong, hot, pithy and unpalatable. If planted in rich, new soil, they will be firm, crisp, juicy and sweet. This fact holds good with Ginseng.

If planted in old ground that is highly fertilized, the roots will grow large, but the flavor is altogether different from that of the wild root, and no doubt specimens of large sizes are spongy and unpalatable to the Celestials compared to that of the wild root.

If planted in rich, new ground and no strong fertilizer used, depending entirely upon the rich woods soil for enriching the beds, the flavor is bound to be exactly as that of the wild root. When the growers wake up to this fact, and dig their roots before they become too large, prices will be very satisfactory and the business will be on a sound basis.

* * *

We will begin in a systematic way, with the location, planting and preparing of the ground for the Ginseng garden, writes a successful grower--C. H. Peterson--of Blue Earth County, Minn.

In choosing a location for a Ginseng garden, select one having a well-drained soil. Ginseng thrives best in wood loam soil that is cool

CHAPTER VI. 32

and mellow, although any good vegetable garden soil will do very well. A southern slope should be avoided, as the ground gets too warm in summer and it also requires more shade than level or northern slope does. It is also apt to sprout too early in the spring, and there is some danger of its getting frosted, as the flower stem freezes very easily and no seed is the result.

Then again if you locate your garden on too low ground the roots are apt to rot and the freezing and thawing of wet ground is hard on Ginseng. Laying out a garden nothing is more important than a good system both for looks, convenience and the growth of your roots later on. Do your work well as there is good money in raising Ginseng, and for your time you will be well repaid. Don't make one bed here and another there and a path where you happen to step, but follow some plan for them. I have found by experience that the wider the beds are, the better, providing that their width does not exceed the distance that you can reach from each path to center of bed to weed. For general purposes for beds 6 1/2 ft. is used for paths 1 1/2 ft. A bed 6 1/2 ft. wide gives you 3 1/4 ft. to reach from each path to center of bed without getting on the beds, which would not be advisable. An 18 in. path is none too wide after a few years' growth, as the plants nearly cover this with foliage. This size beds and paths are just the right width for the system of lath shading I am using, making the combined distance across bed and path 8 ft., or 16 ft. for two beds and two paths, just right to use a 1x4 rough 16 ft. fencing board to run across top of posts described later on.

[Illustration: Plan for Ginseng Garden 24x40 Feet--Ground Plan one line, overhead dotted.]

Now we will lay out the garden by setting a row of posts 8 ft. apart the length you desire to make your garden. Then set another row 8 ft. from first row running parallel with first row, and so on until desired width of your garden has been reached. Be sure to have post line up both ways and start even at ends. Be sure to measure correctly. After all posts are set run a 1x4 in. rough fence board across garden so top edge is even at top of post and nail to post. The post should be about 8 ft. long so when set would be a trifle over 6 ft. above ground. This enables a person to walk under shading when completed. It is also cooler for

CHAPTER VI.

your plants. In setting the posts do not set them too firm, so they can be moved at top enough to make them line up both ways. After the 1x4 in. fence board is put on we will nail on double pieces.

Take a 1x6 rough fence board 16 ft. long and rip it so as to make two strips, one 3 1/2 and the other 2 1/2 inches wide, lay the 3 1/2 in. flat and set the 2 1/2 in. strip on edge in middle of other strip and nail together. This had better be done on the ground so it can be turned over to nail. Then start at one side and run this double piece lengthwise of your garden or crosswise of the 1x4 in. fence board nailed along top of post and nail down into same. It may be necessary to nail a small piece of board on side of the 1x4 in. board where the joints come. Then lay another piece similar to this parallel with first one, leaving about 49 1/2 in. between the two. This space is for the lath panel to rest on the bottom piece of the double piece. Do not put double pieces so close that you will have to crowd the lath panels to get them in, but leave a little room at end of panel. You will gain about 1 1/2 in. for every double piece used in running across the garden. This has to be made up by extending over one side or the other a piece of 1x4 board nailed to end of 1x4 board nailed at top posts. Let this come over the side you need the shade most. Begin from the side you need the shade least and let it extend over the other side.

It is advisable to run paths on outside of garden and extend the shading out over them. On sides lath can be used unless otherwise shaded by trees or vines. It will not be necessary to shade the north side if shading extends out over end of beds several feet. Give your plants all the air you can. In this system of shading I am using I have figured a whole lot to get the most convenient shading as well as a strong, substantial one without the use of needless lumber, which means money in most places. It has given good satisfaction for lath shade so far. Being easily built and handy to put on in spring and take off in fall.

Now don't think I am using all lath shade, as I am not. In one part of garden I am using lath and in another part I am using some good elm trees. I think, however, that the roots make more rapid growth under the lath shade, but the trees are the cheaper as they do not rot and have to be replaced. They also put on their own shade. The leaves

CHAPTER VI.   34

when the proper time comes also removes it when the time comes in the fall and also mulches the beds at the same time.

We will now plan out the beds and paths. Use 1x4 in. rough 16 ft. fence boards on outside row of posts next to ground, nail these to posts, continue and do likewise on next row of posts, and so on until all posts have boards nailed on same side of them as first one, the post being just on inside edge of your beds. Then measure 6 1/2 ft. toward next board, drive a row of stakes and nail a board of same width to same the length of your garden that will make 18 in. between last row of boards and boards on next row nailed to post for the path.

These boards answer several purposes, viz., keep people from walking on beds, elevates beds above paths, holds your mulching of leaves and adds to the appearance of your garden. After beds are made by placing the boards spade the ground about a foot deep all over the bed so as to work it up in good shape. After this is done fork it over with a six-tine fork. If bed is made in summer for fall or spring planting it is well to work it over several times during the summer, as the ground cannot be too mellow. This will also help kill the weeds. Then just before planting rake it down level.

In case beds are made in woods cut, or better, grub out all trees not needed for shade, and if tree roots are not too large cut out all next to the surface running inside of boards in beds, and work the same as other beds. Lay out your beds same as for lath shade with paths between them. Don't try to plant Ginseng in the woods before making it into beds, as you will find it unsatisfactory.

We will now make the lath panel before mentioned.

[Illustration: A Lath Panel.]

Place three laths so that when the laths are laid crosswise one of the laths will be in the middle and the other two, one at each end two inches from end. Can be placed at the end, but will rot sooner. Then begin at end of the three laths and nail lath on, placing them 1/2 in. apart until other end is reached, and if lath is green put closer together to allow for shrinkage. If you have many panels to make, make a table

CHAPTER VI.   35

out of boards and lay strips of iron fastened to table where the three lath comes, so as to clinch nails when they strike the iron strips, which will save a lot of work. Gauges can also be placed on side of table to lay lath so they will be even at ends of panels when finished. Then lay panels in your double pieces on your garden, and if garden is not located in too windy a locality they will not blow out without nailing, and a wire drawn tight from end to end of garden on top of panels will prevent this, and is all that is necessary to hold them in place.

In Central New York, under favorable conditions, Ginseng plants should be coming up the last of April and early May, and should be in the ground by or before April 1st, to give best results. Healthy roots, taken up last of March or early April will be found covered with numerous fine hair-like rootlets. These are the feeders and have all grown from the roots during the spring. They should be well established in the soil before plants appear. Fifteen minutes exposure to the sun or wind will seriously injure and possibly destroy these fine feeders, forcing the roots to throw out a second crop of feeders.

Considering these conditions and frequent late seasons, our advice to beginners is, wait until fall for transplanting roots. But we are not considering southern conditions. Southern growers must be governed by their own experience and climatic conditions. It may be a matter of convenience sometimes for a northern grower to take up one or two year seedlings and transplant to permanent beds in spring. If conditions are favorable so the work can be done in March or early April, it may be allowable. Have ground ready before roots are taken up. Only take up a few at a time, protect from sun and wind, transplant immediately.

Spring sowing of old seed. By this we mean seed that should have been sowed the fall before when one year old, but has been kept over for spring sowing.

[Illustration: One, Two and Three Year Old Ginseng Roots.]

There is other work that can be done quite early in the Ginseng gardens. All weeds that have lived thru the winter should be pulled as soon as frost is out of ground. They can be pulled easier then than any

CHAPTER VI.

other time and more certain of getting the weed root out. Mulching should be looked to. When coarse material like straw or leaves has been used, it should be loosened up so air can get to the soil and the plants can come up thru the mulch. If very heavy, perhaps a portion of the mulch may need to be removed, but don't! don't! take mulch all off from beds of set roots. Seed beds sown last fall will need to be removed about time plants are starting up. But seed beds should have been mulched with coarse leaf loam, or fine vegetable mulch, and well rotted horse manure (half and half), thoroughly mixed together, this mulch should have been put on as soon as seeds were sown and covered with mulch one inch deep. If this was not done last fall it should be put on this spring as soon as snow is off beds.

[Illustration: Ginseng plants "coming up."]

There is another point that needs careful attention when plants are coming up. On heavy soil plants are liable to be earth bound; this is quite likely to occur on old beds that have not been mulched and especially in dry seasons. As the Ginseng stalk comes out of the ground doubled (like an inverted U) the plant end is liable to be held fast by the hard soil, causing injury and often loss of plants. A little experience and careful observation will enable one to detect earth bound plants. The remedy is to loosen soil around the plant. A broken fork tine about eight inches long (straightened) and drive small end in a piece of broom handle about four inches long for a handle, flatten large end of tine like a screwdriver; this makes a handy tool for this work. Force it into soil near plant, give a little prying movement, at same time gently pull on plant end of stalk until you feel it loosen, do not try to pull it out, it will take care of itself when loosened. There is not likely to be any trouble, if leaves appear at the surface of soil. This little spud will be very useful to assist in pulling weed roots, such as dandelion, dock, etc.

Where movable or open shades are used, they need not be put on or closed till plants are well up; about the time leaves are out on trees is the general rule. But one must be governed to some extent by weather and local conditions. If warm and dry, with much sun, get them on early. If wet and cool, keep them off as long as practicable, but be ready to get them on as soon as needed.

CHAPTER VI.

I would advise a would-be grower of Ginseng to visit, if possible, some gardens of other growers and learn all they can by inquiry and observation.

In selecting a place for your garden, be sure it has good drainage, as this one feature may save you a good deal of trouble and loss from "damping off," "wilt," and other fungus diseases which originate from too damp soil.

A light, rich soil is best. My opinion is to get soil from the forest, heap up somewhere for a while thru the summer, then sift thru sand sieve or something similar, and put about two inches on top of beds you have previously prepared by spading and raking. If the soil is a little heavy some old sawdust may be mixed with it to lighten it. The woods dirt is O. K. without using any commercial fertilizers. The use of strong fertilizers and improper drying is responsible for the poor demand for cultivated root. The Chinese must have the "quality" he desires and if flavor of root is poor, will not buy.

* * *

I wonder how many readers know that Ginseng can be grown in the house? writes a New York dealer.

Take a box about 5 inches deep and any size you wish. Fill it with woods dirt or any light, rich soil. Plant roots in fall and set in cellar thru the winter. They will begin to come up about April 1st, and should then be brought out of cellar. I have tried this two seasons. Last year I kept them by a window on the north side so as to be out of the sunshine. Window was raised about one inch to give ventilation. Two plants of medium size gave me about 100 seeds.

This season I have several boxes, and plants are looking well and most of them have seed heads with berries from one-third to three-fourths grown. They have been greatly admired, and I believe I was the first in this section to try growing Ginseng as a house plant.

* * *

CHAPTER VI.

As to the location for a Ginseng garden, I have for the past two years been an enthusiast for cultivation in the natural forest, writes L. C. Ingram, M. D., of Minnesota. It is true that the largest and finest roots I have seen were grown in gardens under lattice, and maintaining such a garden must be taken into account when balancing your accounts for the purpose of determining the net profits, for it is really the profits we are looking for.

The soil I have found to be the best, is a rich black, having a good drain, that is somewhat rolling. As to the direction of this slope I am not particular so long as there is a rich soil, plenty of shade and mulch covering the beds.

The selection of seed and roots for planting is the most important item confronting the beginner. Considerable has been said in the past concerning the distribution among growers of Japanese seed by unscrupulous seed venders. It is a fact that Japanese Ginseng seed have been started in a number of gardens, and unless successfully stamped out before any quantity finds its way into the Chinese market, the Ginseng industry in America, stands in peril of being completely destroyed. Should they find our root mixed, their confidence would be lost and our market lost. Every one growing Ginseng must be interested in this vital point, and if they are suspicious of any of their roots being Japanese, have them passed upon by an expert, and if Japanese, every one dug.

[Illustration: Bed of 10,000 Young Ginseng Plants in Forest.]

It is a fact that neighboring gardens are in danger of being mixed, as the bees are able to do this in carrying the mixing pollen. The safest way to make a start is by procuring seed and roots from the woods wild in your own locality. If this cannot be done then the seed and roots for a start should be procured from a reliable party near you who can positively guarantee the seed and roots to be genuine American Ginseng. We should not be too impatient and hasty to extend the garden or launch out in a great way. Learn first, then increase as the growth of new seed will permit.

CHAPTER VI.

The next essential thing is the proper preparation of the soil for the planting of the seeds and roots. The soil must be dug deep and worked perfectly loose same as any bed in a vegetable garden. The beds are made four or five feet wide and raised four to six inches above the paths, which are left one and a half to two feet wide. I have had seed sown on the ground and covered with dirt growing beside seed planted in well made beds and the contrast in size and the thriftiness of roots are so great when seen, never to be forgotten. The seedlings growing in the hard ground were the size of oat kernels, those in the beds beside them three to nine inches long and weighing from four to ten times as much per root.

In planting the seed all that is necessary is to scatter the stratified seed on top of the prepared bed so they will be one or two inches apart, then cover with loose dirt from the next bed then level with back of garden rake. They should be one-half to one inch covered. Sawdust or leaves should next be put on one to two inches for a top dressing to preserve moisture, regulate heat, and prevent the rains from packing the soil.

The best time to do all planting is in the spring. This gives the most thrifty plants with the least number missing. When the plants are two years old they must be transplanted into permanent beds. These are prepared in the same manner as they were for the seed. A board six inches wide is thrown across the bed, you step on this and with a spade throw out a ditch along the edge of the board. In this the roots are set on a slant of 45 degrees and so the bud will be from one to two inches beneath the surface. The furrow is then filled and the board moved its width. By putting the roots six inches apart in the row and using a six-inch board your plants will be six inches each way, which with most growers have given best results. When the roots have grown three years in the transplanted beds they should be ready to dig and dry for market. They should average two ounces each at this time if the soil was rich in plant food and properly prepared and cared for.

The plants require considerable care and attention thru each summer. Moles must be caught, blight and other diseases treated and the weeds pulled, especially from among the younger plants. As soon as the plants are up in the spring the seed buds should be clipped from all the plants except those finest and healthiest plants you may save for your seed to maintain your garden. The clipping of the seed buds is very essential, because we want the very largest and best flavored root in the shortest time for the market. Then if we grow bushels of seed to the expense of the root, it is only a short time when many thousands of pounds of root must compete with our own for the market and lower the price.

CHAPTER VII.

SHADING AND BLIGHT.

In several years experience growing Ginseng, says a well known grower, I have had no trouble from blight when I shade and mulch enough to keep the soil properly cool, or below 65 degrees, as you will find the temperature in the forests, where the wild plants grow best, even during summer days.

Some years ago I allowed the soil to get too warm, reaching 70 degrees or more. The blight attacked many plants then. This proved to me that growing the plants under the proper temperature has much to do with blight.

When fungus diseases get upon wild plants, that is plants growing in the forest, in most cases it can be traced to openings, forest fires and the woodman's ax. This allows too much sun to strike the plants and ground in which they are growing. If those engaged, or about to engage, in Ginseng growing will study closely the conditions under which the wild plants flourish best, they can learn much that they will only find out after years of experimenting.

Mr. L. E. Turner in a recent issue of "Special Crops" says: We cannot depend on shade alone to keep the temperature of the soil below 65 degrees--the shade would have to be almost total. In order to allow sufficient light and yet keep the temperature down, we must cover the ground with a little mulch. The more thoroughly the light is diffused the better for the plants. Now, when we combine sufficient light with say one-half inch of clean mulch, we are supplying to the plants their natural environment, made more perfect in that it is everywhere alike.

The mulch is as essential to the healthy growth of the Ginseng plant as clothing is to the comfort and welfare of man; it can thrive without it no more than corn will grow well with it. These are plants of opposite nature. Use the mulch and reduce the shade to the proper density. The mulch is of the first importance, for the plants will do much better with the mulch and little shade than without mulch and with plenty of shade.

CHAPTER VII.

Ginseng is truly and wholly a savage. We can no more tame it than we can the partridge. We can lay out a preserve and stock it with Ginseng as we would with partridges, but who would stock a city park with partridges and expect them to remain there? We cannot make a proper Ginseng preserve under conditions halfway between a potato patch and a wild forest, but this is exactly the trouble with a large share of Ginseng gardens. They are just a little too much like the potato patch to be exactly suited to the nature of Ginseng. The plant cannot thrive and remain perfectly healthy under these conditions; we may apply emulsions and physic, but we will find it to be just like a person with an undermined constitution, it will linger along for a time subject to every disease that is in the air and at last some new and more subtle malady will, in spite of our efforts, close its earthly career.

Kind readers, I am in a position to know thoroughly whereof I write, for I have been intimate for many years with the wild plants and with every shade of condition under which they manage to exist. I have found them in the valley and at the hilltop, in the tall timber and the brambled "slashing," but in each place were the necessary conditions of shade and mulch. The experienced Ginseng hunter comes to know by a kind of instinct just where he will find the plant and he does not waste time searching in unprofitable places. It is because he understands its environment. It is the environment he seeks--the Ginseng is then already found. The happy medium of condition under which it thrives best in the wild state form the process of healthy culture.

[Illustration: One Year's Growth of Ginseng Under Lattice Shade.]

Mr. Wm E. Mowrer, of Missouri, is evidently not in favor of the cloth shading. I think if he had thoroughly water-proofed the cloth it would have withstood the action of the weather much better. It would have admitted considerably less light and if he had given enough mulch to keep the soil properly cool and allowed space enough for ventilation, he would not have found the method so disastrous. We will not liken his trial to the potato patch, but to the field where tobacco is started under canvas. A tent is a cool place if it is open at the sides and has openings in the top and the larger the tent the cooler it will be. Ginseng does splendidly under a tent if the tent is built expressly with regard to the requirements of Ginseng.

CHAPTER VII.

In point of cheapness a vine shading is yet ahead of the cloth system. The wild cucumber vine is best for this purpose, for it is exactly suited by nature to the conditions in a Ginseng garden. It is a native of moist, shady places, starts early, climbs high and rapidly. The seeds may be planted five or six in a "hill" in the middle of the beds, if preferred, at intervals of six or seven feet, and the vines may be trained up a small pole to the arbor frame. Wires, strings or boughs may be laid over the arbor frame for the vines to spread over. If the shade becomes too dense some of the vines may be clipped off and will soon wither away. Another advantage of the wild cucumber is that it is very succulent, taking an abundance of moisture and to a great extent guards against excessive dampness in the garden. The vines take almost no strength from the soil. The exceeding cheapness of this method is the great point in its favor. It is better to plant a few too many seeds than not enough, for it is easy to reduce the shade if too dense, but difficult to increase it in the summer if too light.

* * *

This disease threatens seriously to handicap us in the raising of Ginseng, says a writer in "Special Crops." It does down, but is giving us trouble all over the country. No section seems to be immune from it, tho all seem to be spraying more or less. I know of several good growers whose gardens have gone down during the last season and this, and they state that they began early and sprayed late, but to no decided benefit. What are we to do? Some claim to have perfect success with spraying as their supposed prevention.

Three years ago I began to reason on this subject and in my rambles in the woods, I have watched carefully for this disease, as well as others on the wild plant, and while I have now and then noted a wild plant that was not entirely healthy, I have never seen any evidence of blight or other real serious disease. The wild plant usually appears ideally healthy, and while they are smaller than we grow in our gardens, they are generally strikingly healthful in color and general appearance. Why is this so? And why do we have such a reverse of things among our gardens?

CHAPTER VII.

I will offer my ideas on the subject and give my theories of the causes of the various diseases and believe that they are correct and time will prove it. At least I hope these efforts of mine will be the means of helping some who are having so much trouble in the cultivation of Ginseng. The old saw that the "proof of the pudding is in chewing the bag," may be amply verified by a visit to my gardens to show how well my theories have worked so far. I will show you Ginseng growing in its highest state of perfection and not a scintilla of blight or any species of alternaria in either of them, while around me I scarcely know of another healthy garden.

To begin with, moisture is our greatest enemy; heat next; the two combined at the same time forming the chief cause for most diseases of the plant.

If the soil in our gardens could be kept only slightly moist, as it is in the woods, and properly shaded, ventilated and mulched, I am sure such a thing as blight and kindred diseases would never be known. The reason for this lies in the fact that soil temperature is kept low and dry. The roots, as is well known, go away down in the soil, because the temperature lower down is cooler than at the surface.

Here is where mulch plays so important a part because it protects the roots from so much heat that finds its way between the plants to the top of the beds. The mulch acts as a blanket in keeping the heat out and protecting the roots thereby. If any one doubts this, just try to raise the plants without mulch, and note how some disease will make its appearance. The plant will stand considerable sun, however, with heavy enough mulch. And the more sun it can take without harm, the better the root growth will be. Too much shade will show in a spindling top and slender leaves, and invariable smallness of root growth, for, let it be borne in mind always, that the plant must derive more or less food from the top, and it is here that the fungi in numerous forms proceed to attack.

The plant will not grow in any other atmosphere but one surcharged with all kinds of fungi. This is the natural environment of the plant and the only reason why the plants do not all become diseased lies in the plain fact that its vitality is of such a high character that it can resist the disease, hence the main thing in fighting disease is to obtain for the plant the best possible hygienic surroundings and feed it with the best possible food and thus nourish it to the highest vitality.

I am a firm believer in spraying of the proper kind, but spraying will not keep a plant free from disease with other important conditions lacking. Spraying, if heavily applied, is known as a positive injury to the plant, despite the fact that many claim it is not, and the pity is we should have to resort to it in self-defense. The pores of the leaflets are clogged up to a greater or less extent with the deposited solution and the plant is dependent to this extent of its power to breathe.

Coat a few plants very heavily with spray early in the season and keep it on and note how the plants struggle thru the middle of a hot day to get their breath. Note that they have a sluggish appearance and are inclined to wilt. These plants are weakened to a great extent and if an excess of moisture and heat can get to them, they will perhaps die down. Another thing: Take a plant that is having a hard time to get along and disturb the root to some extent and in a day or two notice spots come upon it and the leaves begin to show a wilting. Vitality disturbed again.

[Illustration: A Healthy Looking Ginseng Garden.]

The finest plants I have ever found in the woods were growing about old logs and stumps, where the soil was heavily enriched with decaying wood. A good cool spot, generally, and more or less mulch, and if not too much shade present. Where the shade was too dense the roots were always small. I have in some instances found some very fine roots growing in the midst of an old stump with no other soil save the partially rotted stump dirt, showing thus that Ginseng likes decaying wood matter. Upon learning this, I obtained several loads of old rotten sawdust, preferably white oak or hickory and my bed in my gardens is covered at least two inches with it under the leaf mulch. This acts as a mulch and natural food at one and the same time. The leaves decay next to the soil and thus we supply leaf mold.

This leaf mold is a natural requirement of the plant and feeds it also constantly. A few more leaves added each fall keep up the process

CHAPTER VII.

and in this way we are keeping the plant wild, which we must do to succeed with it, for Ginseng can not be greatly changed from its nature without suffering the consequences. This is what is the matter now with so many of us. Let's go back to nature and stay there, and disease will not give us so much trouble again.

One more chief item I forgot to mention was the crowding of the plants together. The smaller plants get down under the larger and more vigorous and have a hard struggle for existence. The roots do not make much progress under these conditions, and these plants might as well not be left in the beds. And also note that under those conditions the beds are badly ventilated and if any plants are found to be sickly they will be these kind. I shall plant all my roots henceforth at least ten inches apart each way and give them more room for ventilation and nourishment. They get more chance to grow and will undoubtedly make firm root development and pay largely better in the end. Corn cannot be successfully cultivated in rows much narrower than four feet apart and about two stalks to the hill. All farmers know if the hills are closer and more stalks to the hill the yield will be much less.

At this point I would digress to call attention to the smallness of root development in the woods, either wild or cultivated, because the trees and tree roots sap so much substance from the soil and other weeds and plants help to do the same thing. The shade is not of the right sort, too dense or too sparse in places, and the plants do not make quick growth enough to justify the growing under such conditions, and while supposed to be better for health of plants, does not always prove to be the case. I have seen some gardens under forest shade that blighted as badly as any gardens.

So many speak of removing the leaves and mulch in the spring from the beds. Now, this is absolutely wrong, because the mulch and leaves keep the ground from becoming packed by rains, preserves an even moisture thru the dry part of the season and equalizes the temperature. Temperature is as important as shade and the plants will do better with plenty of mulch and leaves on the beds and considerable sun than with no mulch, dry hard beds and the ideal shade. Roots make but little growth in dry, hard ground. Pull your weeds out by hand and protect your garden from the seng digger thru the summer and that will be your cultivation until September or October when you must transplant your young roots into permanent beds, dig and dry the mature roots.

CHAPTER VIII.

DISEASES OF GINSENG.

The following is from an article on "The Alternaria Blight of Ginseng" by H. H. Whetzel, of Cornell University, showing that the author is familiar with the subject:

Susceptibility of Ginseng to Disease.

The pioneer growers of Ginseng thought they had struck a "bonanza." Here was a plant that seemed easily grown, required little attention after it was once planted, was apparently free from all diseases to which cultivated plants are heir and was, besides, extremely valuable. Their first few crops bore out this supposition. No wonder that a "Ginseng craze" broke out and that men sat up nights to figure out on paper the vast fortunes that were bound to accrue to those who planted a few hundred seeds at three cents each and sold the roots in five years at $12.00 a pound.

Like many other grow-wealthy-while-you-wait schemes, nature herself imposed a veto. Diseases began to appear. The prospective fortune shrunk, frequently dried up and blew away or rotted and disappeared in the earth. Several factors contributed to this result:

1. The removal of a wild plant from its natural habitat to an entirely artificial one.

2. The encouragement by the application of manures and cultivation of a rapidity of growth to which the plant was by inheritance an entire stranger, thus weakening its constitution and depriving it of its natural ability to withstand disease. Cultivated roots in three years from the seed attain greater size than they often would in twenty years in the woods.

3. The failure in many cases to provide conditions in any degree approximating the natural habitat, as, for example, the failure to supply proper drainage that is in nature provided by the forest trees whose roots constantly remove the excess of rainfall.

CHAPTER VIII.

[Illustration: Diseased Ginseng Plants.]

4. The crowding of a large number of plants into a small area. This, in itself, is more responsible for disease epidemics than perhaps any other factor.

Of all the twelve or fifteen, now more or less known, diseases of this plant one in particular stands out as *the disease* of Ginseng. Altho one of the latest to make its appearance, it has in three or four years spread to nearly every garden in this state and its ravages have been most severe. This disease is the well known Alternaria Blight.

The Most Common and Destructive Disease of Ginseng.

The disease manifests itself in such a variety of ways, depending upon the parts of the plant attacked, that it is difficult to give a description by which it may always be identified. It is usually the spotting of the foliage that first attracts the grower's attention. If examined early in the morning the diseased spots are of a darker green color and watery as if scalded. They dry rapidly, becoming papery and of a light brown color, definite in outline and very brittle. With the return of moist conditions at night the disease spreads from the margin of the spot into the healthy tissue. The disease progresses rapidly so that in a very few days the entire leaf succumbs, wilts and hangs limp from the stalk. If the weather is wet, the progress of the disease is often astonishing, an entire garden going down in a day or two. Under such conditions the leaves may show few or no spots becoming thruout of a dark watery green and drooping as if dashed with scalding water. All parts of the top may be affected. The disease never reaches the roots, affecting them only indirectly.

Cause of the Disease.

The disease is the result of the growth of a parasitic fungus in the tissues of the Ginseng. This fungus is an Alternaria (species not yet determined) as is at once evident from an examination of its spores. These are in size and form much like those of the early Blight Alternaria of Potato. These spores falling upon any part of the plant above the ground will, if moisture be present, germinate very quickly, sending out germ tubes which pierce the epidermis of the host. These mycelium threads ramify thru the tissues of the leaf or stem as the case may be, causing death of the cells. From the mycelium that lies near or on the surface arise clusters or short brown stalks or conidiophores on the apex of which the spores are borne in short chains. The spores mature quickly and are scattered to healthy plants, resulting in new infections. Only one form of spores, the conidial, is at present known.

That the Alternaria is a true parasite and the cause of the disease there can be no doubt. The fungus is constantly associated with the disease. Inoculation experiments carried on in the botanical laboratory this summer show conclusively that the germ tube of the spore can penetrate the epidermis of healthy Ginseng leaves and stems and by its growth in such healthy tissue cause the characteristic spots of the disease. This is of special interest as it adds another to the list of parasitic species of genus long supposed to contain only saprophytes.

Upon the general appearance of so destructive a disease, one of the first questions of the growers was "where did it come from?" Believing that it was a natural enemy of the wild plant, now grown over powerful under conditions highly unnatural to Ginseng, I undertook to find proof of my theory. I visited a wooded hillside where wild Ginseng was still known to exist. After half a day's diligent search I obtained seventeen plants of different ages, one of which showed spots of the Blight. Examination with the microscope showed mycelium and spores of the Alternaria. Unfortunately I did not get pure cultures of the fungus from this plant and so could not by cross inoculations demonstrate absolutely the identity of the Alternaria on the wild plant with that of the cultivated. So far, however, as character of the spots on the leaves, size and form of the spores are concerned, they are the same. This, I believe, answers the question of the source of the disease. Introduced into gardens on wild plants brought from the woods, it has spread rapidly under conditions most favorable to its development; namely, those pointed out in the earlier part of this paper.

The wind, I believe, is chiefly responsible for the dissemination of the spores which are very small and light. Not only does the wind carry the spores from plant to plant thruout the garden, but no doubt frequently

CHAPTER VIII.   51

carries them for longer distances to gardens near by. The spores are produced most abundantly under conditions favorable to such dissemination. During moist, cloudy weather the energies of the fungus are devoted to vegetative growth, the spreading of the mycelium in the host tissues. With the advent of bright sunny days and dry weather mycelium growth is checked and spore formation goes on rapidly. These spores are distributed when dry and retain their vitality for a long period. Spores from dried specimens in the laboratory have been found to germinate after several months when placed in water. The disease might also be very readily carried by spores clinging to the roots or seeds, or possibly even by the mycelium in the seeds themselves. The fungus very probably winters in the old leaves and stems or in the mulch, living as a saprophyte and producing early in the spring a crop of spores from which the first infections occur.

Summer History of the Disease.

Altho it is on the foliage that the disease first attracts the attention of the grower, it is not here that it really makes its first appearance in the spring. The stem is the first part of the plant to come thru the soil and it is the stem that is first affected. The disease begins to show on the stems very shortly after they are thru the soil, evident first as a rusty, yellow spot usually a short distance above the surface of the soil or mulch. The spot rapidly increases in size, becomes brown and finally nearly black from the multitude of spores produced on its surface. The tissue of the stem at the point of attack is killed and shrinks, making a canker or rotten strip up the side of the stem. Such stems show well developed leaves and blossom heads giving no evidence of the disease beneath. Occasionally, however, the fungus weakens the stem so that it breaks over. Growers have occasionally observed this "stem rot" but have never connected it with the disease on the leaves later in the season.

[Illustration: Broken--"Stem Rot."]

It is from the spores produced on these cankers on the stem that the leaves become infected. The disease begins to appear on the leaves some time in July and by the middle of August there is usually little foliage alive. Infection frequently occurs at the point where the five leaflets are attached to the common petiole. The short leaf stems are killed causing the otherwise healthy leaflets to droop and wilt. This manifestation of the disease has not generally been attributed to the Alternaria. The seedlings are frequently affected in the same way causing what is sometimes known as the "top blight of seedlings."

From the diseased leaves and stems the spores of the fungus find their way to the seed heads which at this time are rapidly filling out by the growth of the berries. The compact seed heads readily retain moisture, furnishing most favorable conditions for the germination of any spores that find their way into the center of the head. That this is the usual course of seed head infection is shown by the fact that it is the base of the berry on which the spots start. These spots, of a rusty yellow color, gradually spread all over the seed which finally becomes shriveled and of a dark brown or black color. Spores in abundance are formed on the diseased berries. Affected berries "shell" from the head at the slightest touch. This manifestation of the disease has long been known as "seed blast." If the berries have begun to color the injury from the disease will probably be very slight. The "blasting" of the green berries, however, will undoubtedly reduce or destroy the vitality of the seed. There is a strong probability that the fungus may be carried over in or on the seed.

[Illustration: End Root Rot of Seedlings.]

The roots are only indirectly affected by this disease. The fungus never penetrates to them. Roots from diseased tops will grow perfectly normal and healthy plants the following season. It is in the leaves of the plant that practically all of the substance of the root is made. The bulk of this substance is starch. The destruction of the foliage, the manufacturing part of the plant, long before it would normally die means of course some reduction in the growth and starch content of the root. However, it seems probable that the greater portion of root growth is made before the blight attacks the foliage. This seems borne out by the fact that even blighted seedlings usually show nearly as good growth and bud development as those not blighted. In the case of older plants this is probably much more true as the latter part of the season is devoted largely to growing and maturing the berries. The Alternaria blight is dreaded chiefly because of its destructive effects on

CHAPTER VIII.

the seed crop.

Preventive.

The first experimental work on the control of this disease so far as I know, was carried out by Dr. I. C. Curtis of Fulton, N. Y. Having suffered the total loss of foliage and seed crop during the season of 1904, Dr. Curtis determined to test the efficacy of the Bordeaux mixture the following season as a preventive of the blight. The success of his work, together with this method of making and applying the mixture is given by him in Special Crops for January, 1906.

Extensive experiments in spraying were carried out during the past season by the Ginseng Company at Rose Hill, N. Y., under the direction of the writer. During 1905 their entire seed crop was completely destroyed by the blight. Losses from the same disease the previous season had been very heavy. During 1905 they had succeeded in saving a very large proportion of their seedlings by spraying them with the Bordeaux mixture. Encouraged by this they began spraying early in the spring of 1906, just when the plant began to come thru the ground. This was repeated nearly every week during the season, the entire ten acres being sprayed each time. On account of poor equipment the earlier sprayings were not as thoroughly done as they should have been, and some disease appeared on the stalks here and there thruout the gardens. A new pump and nozzles were soon installed and all parts of the plant completely covered. Practically no blight ever appeared on the foliage. There was some loss from "blast of seed heads" due to a failure to spray the seed heads thoroughly while they were filling out. The seed heads Were doubtless infected from the diseased stalks that had not been removed from the garden. A very large seed crop was harvested. The formula of the Bordeaux used at Rose Hill was about 4-6-40, to each one hundred gallons of which was added a "sticker" made as follows:

Two pounds resin. One pound sal soda (Crystals). One gallon water.

Boiled together in an iron kettle until of a clear brown color. It is probable that more applications of Bordeaux were given than was necessary, especially during the middle part of the season when little new growth was being made.

From these experiments it is evident that the problem of the control of the Alternaria Blight of Ginseng has been solved. Thorough spraying with Bordeaux mixture begun when the plants first come thru the ground and repeated often enough to keep all new growths covered, will insure immunity from the blight. Thoroughness is the chief factor in the success of this treatment. It is, however, useless to begin spraying after the disease has begun to appear on the foliage.

* * *

To the President and Members of the Missouri State Ginseng Growers' Association.

Gentlemen--In response to a request from your secretary, I was sent early in August to investigate your Ginseng gardens, and, if possible, to give some help in checking a destructive disease which had recently appeared and had in a short time ruined much of the crop. Thru the aid of some of your association, at the time of my visit to Houston, and since that time, I have been furnished with valuable data and specimens of diseased plants.

The summer of 1904 was marked by a very abundant rainfall. The shade of the arbors kept the soil beneath them moist, if not wet, for several weeks at a time. This moist soil, rich in humus and other organic substances, formed an exceedingly favorable place for the growth of fungi. Gardens under dense shade with poor drainage, suffered the greatest loss. All ages of plants were attacked and seemed to suffer alike, if the conditions were favorable for the growth of fungi.

Symptoms of Disease and Nature of the Injury.

Between the first and the fifteenth of May black spots having the appearance of scars appeared on the stems of the Ginseng plants. All ages of plants were attacked. The scars increased in number and grew in size, sometimes encircling the stem.

CHAPTER VIII.

The first indication of injury was seen when one leaflet after another turned brown; from them the disease spread down the petiole to the main stalk. Other stalks were attacked so badly that they broke off and fell over before the upper portions had even become withered. After the loss of the top from this disease the crown of the root was liable to be attacked by fungi or bacteria, causing decay. I found little of this in the gardens at Houston. The greatest loss caused by this disease lies in the destruction of the seed crop.

I have succeeded in isolating and studying the fungus which causes this disease. The fungus belongs to the genus Vermicularia and occurs on a number of our common herbaceous plants. I found it near Columbia this autumn on the Indian turnip. The fungus lives beneath the epidermis of the Ginseng plant; breaking the epidermis to form the black scars in which the spores, or reproductive bodies, are produced. The spores when ripe are capable of germinating and infecting other plants.

Treatment.

Fortunately this disease can be effectually checked by the use of Bordeaux spraying mixture.

Damping-off Disease.

Another source of loss was in the damping-off of young plants. The fungus which causes this disease lives in the surface layer of the soil and girdles the plants at the surface layer of the ground, causing them to wilt and fall over. The trouble can be largely avoided by proper drainage and stirring the surface layer, thus aerating and drying the soil.

The Wilt Disease.

By far the most destructive and dangerous disease remains to be described. It made its appearance about the first week in July, causing the leaves to turn yellow and dry up; the seed stem and berries also dried up and died before reaching maturity. This was the disease which caused the greatest loss; whole plantations often being

CHAPTER VIII.

destroyed in a week. Neither the Bordeaux spraying mixture nor lime dust seemed to check its ravages.

I have succeeded in isolating the fungus which is the cause of this destructive disease and have grown it in the laboratory in pure cultures for nearly five months. Cultures were made by scraping the dark spots on diseased stems with a sterile needle and inoculating sterilized bean pods or plugs of potato with the spores scraped from the stem. In two or three days a white, fluffy growth appears on the bean pod which rapidly spreads until it is covered with a growth which resembles a luxuriant mould. I have also isolated this fungus and made cultures from the soil taken from diseased beds.

The fungus belongs to the genus Fusarium and is probably identical with the fungus which is so destructive in causing the wilt of cotton, watermelon and cowpeas, and which has been carefully studied by Smith and Orton of the United States Department of Agriculture.

Treatment.

It will be seen from this brief description of the fungus that it is an exceedingly difficult disease to combat. Living from year to year in the soil it enters the plants thru the roots and spreads upward thru the water-conducting channels. It does not once appear on the surface until the plant is beyond recovery. Obviously we cannot apply any substance to kill the fungus without first killing the plant it infests.

There is but one conclusion to be drawn, viz.: That application of fungicides will not prevent the wilt disease.

There are, however, two methods of procedure in combating the disease: First, the use of precautions against allowing the fungus to get started; second, the selection and breeding of varieties which will withstand the disease.

From the very first the arbor should be kept free from all possible infection by the wilt fungus.

CHAPTER VIII.

Gardens should be small and located some little distance apart, then if one becomes infected with the disease it can be taken up before the disease infests a larger territory. If the roots have reached merchantable size they had best be dried and sold, since they are likely to carry the disease when transplanted. If they are transplanted they should be carefully cleaned and reset without bruising.

Proper drainage is very necessary for a successful Ginseng garden. It is advisable to locate the garden on a gentle slope if possible. In all cases the ground should be well drained.

The belief of many that the death of the Ginseng was due to the wet season was without foundation, because the fungus develops best in soil which is continually moist and shady. This also accounts for the well-known fact that all rots, mildews and rusts are worse in a rainy season than in a dry one.

[Illustration: The Beginning of Soft Rot.]

Ample ventilation must also be provided in building the arbor. Many arbors are enclosed at the sides too tightly.

The material used for mulching should be of a sort which will not contaminate the garden with disease. Some fungi will be killed if the ground is allowed to freeze before putting on the mulch.

The second and, to my mind, most promising mode of procedure lies in propagating a variety of Ginseng which will be resistant to the wilt disease. In every garden, no matter how badly diseased, there are certain plants which live thru the attacks of the disease and ripen seeds. These seeds should be saved and planted separately, the hardiest of their offspring should be used to propagate seeds for future planting. By thus selecting the hardiest individuals year after year it will be possible in time to originate a variety of parasitic fungi. There seems to me to be more hope in developing such a resistant variety of Ginseng than in discovering some fungicide to keep the disease in check.

Bordeaux Mixture.

CHAPTER VIII.

It is surprising that any considerable number of farmers, horticulturists, Ginseng growers, etc., are ignorant of a preparation so necessary as Bordeaux for profitable cultivation of many crops. The following is taken from Bulletin 194 of the New Jersey Agricultural Experiment Station. The advice given in this paper recently by Professor Craig is repeated and emphasized. Every farmer should have the bulletins issued by the experiment station of his own state and have them within easy reach at all times.

Bordeaux mixture derives its name from the place of its discovery, Bordeaux, France. It consists of copper sulfate, which is commonly called blue vitriol or bluestone, fresh lime and water.

Formulas used--Several strengths of the mixture are used under different conditions:

1. (2:4:50) Copper Sulfate 2 lbs. Quick Lime 4 " Water 50 gals.

2. (3:6:50) Copper Sulfate 3 lbs. Quick Lime 6 " Water 50 gals.

3. (4:4:50) Copper Sulfate 4 lbs. Quick Lime 4 " Water 50 gals.

4. (6:6:50) Copper Sulfate 6 lbs. Quick Lime 6 " Water 50 gals.

Formula 1 is used for very tender foliage, as peach, plum, greenhouse plants, tender seedlings, etc.

Formula 2 which is a half stronger than the preceding has about the same use but for slightly less tender leaves.

Formula 3 is the formula for general use on apples, pears, asparagus, grapes, tomatoes, melons, strawberries, etc.

Formula 4 is the strongest formula that is often used. It is considered best for potatoes and cranberries. It may be used on grapes, on apples and pears before blossoming and sometimes on other crops. It was once more commonly used, but, except as here quoted, it is generally being displaced by Formula 3.

CHAPTER VIII.

* * *

Normal or 1.6 per cent. Bordeaux mixture:

Copper-sulfate (Blue Vitriol) 6 pounds Quick-lime (Good stone lime) 4 " Water 50 gallons

Six pounds of sulfate of copper dissolved in fifty gallons of water, when applied at the proper time, will prevent the growth of fungi. However, if applied in this form, the solution will burn the foliage. Four pounds of quick-lime to six pounds of copper will neutralize the caustic action. When sulfate of copper and lime are added in this proportion, the compound is Bordeaux mixture.

Weighing of copper and lime at time of mixing is very inconvenient. Bordeaux mixture is best when used within a few hours after being mixed. Therefore a stock mixture of Bordeaux is impracticable. It is, however, practicable to have stock preparation of sulfate of copper and of lime ready for mixing when required.

The lime should be fresh quick-lime and when slaked must always be covered with water to exclude the air. In this manner a "stock" mixture of lime can be kept all summer unimpaired.

Sulfate of copper can be dissolved in water and held in solution until needed. One gallon of water will hold in solution two pounds of copper sulfate. To accomplish this the sulfate should be suspended at the surface of the water in a bag. The water most loaded with copper will sink to the bottom and the water least loaded will rise to the surface. If fifty pounds of sulfate are suspended in twenty-five gallons of water on an evening, each gallon of water will, when stirred the next morning, hold two pounds of sulfate. This will form the stock solution of copper sulfate.

If three gallons of this solution are put in the spray barrel, it is equivalent to six pounds of copper. Now fill the spray barrel half full of water before adding any lime. This is important for if the lime is added to so strong a solution of sulfate of copper, a curdling process will follow. Stir the water in the lime barrel so as to make a dilute milk of lime, but never allow it to be dense enough to be of a creamy thickness. If of the latter condition, lumps of lime will clog the spray nozzle. Continue to add to the mixture this milk of lime so long as drops of ferrocyanide of potassium (yellow prussiate of potash) applied to the Bordeaux mixture continue to change from yellow to brown color. When no change of color is shown, add another pail of milk of lime to make the necessary amount of lime a sure thing. A considerable excess of lime does no harm. The barrel can now be filled with water and the Bordeaux mixture is ready for use.

The preparation of ferrocyanide of potassium for this test may be explained. As bought at the drug store, it is a yellow crystal and is easily soluble in water. Ten cents worth will do for a season's spraying of an average orchard. It should be a full saturation; that is, use only enough water to dissolve all the crystals. The cork should be notched or a quill inserted so that the contents will come out in drops. A drop will give as reliable a test as a spoonful. The bottle should be marked "Poison." Dip out a little of the Bordeaux mixture in a cup or saucer and drop the ferrocyanide on it. So long as the drops turn yellow or brown on striking the mixture, the mixture has not received enough lime.

"Process" Lime for Bordeaux Mixture.

The so-called "new process," or prepared limes, now offered on the market, are of two classes. One consists of the quick-lime that has been ground to a powder. The other is the dry water-slaked lime made by using only enough water to slake the quick-lime, but not enough to leave it wet. Practically all of the process lime on the market is the ground quick-lime.

When the hard "stone" lime becomes air-slaked it is evident to the eye from the change to a loose powdery mass. Should one of these prepared limes be to any considerable degree air-slaked, its appearance would be no indication of its real condition.

A simple test for the presence of much carbonate of lime in these prepared limes, can be easily performed, a small amount of lime--1/4 teaspoonful--dropped on a little hot vinegar, will effervesce or "sizzle" if it contain the carbonate of lime, acting about the same as soda.

CHAPTER VIII.

A sample of a new process lime analyzed at this Station showed 30 per cent, magnesia. This came from burning a dolomitic limestone, that is, one containing carbonate of magnesia with the carbonate of lime. The magnesia does not slake with water like the lime and hence is useless in the Bordeaux mixture. There is no easy way outside a chemical laboratory of telling the presence of magnesia.

As a general rule more "process" lime is required to neutralize the copper sulfate than good stone lime. It is always well to make Bordeaux mixture by using the ferrocyanide of potassium test--Cornell University.

CHAPTER IX.

MARKETING AND PRICES.

Preparing Dry Root for Market--There are more growers of Ginseng, I believe, according to Special Crops, who are not fully posted on handling Ginseng root after it is harvested than there are who fail at any point in growing it, unless it may be in the matter of spraying.

There are still many growers who have never dried any roots, and of course know nothing more than has been told them. Stanton, Crossley and others of the pioneers state freely in their writings that three pounds of green root (fall dug) would make one pound of dry.

The market does not want a light, corky, spongy root, neither does it want a root that, when dried, will weigh like a stone. Root when offered to a dealer should be absolutely dry, not even any moisture in the center of the root. Root that is absolutely dry will, in warm, damp weather, collect moisture enough so it will have to be given a day's sun bath or subjected to artificial heat. A root should be so dry that it will not bend. A root the size of a lead pencil should break short like a piece of glass. You ask why this special care to have Ginseng root dry to the last particle of moisture more than any other root. The answer is that Ginseng has to cross the ocean and to insure against its getting musty when sealed up to keep it from the air, it must be perfectly dry.

We know a great many growers have felt hurt because a dealer docked them for moisture, but they should put themselves in the dealer's place. When he disposes of the root it must be perfectly dry. At from $5.00 to $10.00 per pound moisture is rather expensive. The grower should see to it that his root is dry and then instruct the man he ships to that you will stand no cutting.

[Illustration: Dug and Dried--Ready for Market.]

One other cause of trouble between grower and dealer is fiber root. This light, fine stuff is almost universally bought and sold at $1.00 per pound. This seems to be the only stationary thing about Ginseng. It would seem that the fine root could be used in this country for Ginseng

CHAPTER IX.

tincture, but it is not so strong as the regular root, and our chemists prefer the large cultivated root at $5.00 to $7.00 a pound. Now, when your Ginseng root is "dry as a bone," stir it around or handle it over two or three times, and in doing so you will knock off all the little, fine roots. This is what goes in the market as fiber root and should be gathered and put in a separate package. As I said before this fiber root is worth $1.00 per pound and usually passes right along year after year at that same price.

Now as to color. It is impossible to tell just now what color the market will demand. We advise medium. We do not think the extreme dark will be as much sought for as formerly; neither do we think the snow white will be in demand. Now, you can give your Ginseng any color you desire. If you want to dry it white, wash it thoroughly as soon as you dig it. This does not mean two or three hours after being dug, but wash it at once. If you want a very dark root, dig it and spread on some floor and leave it as long as you can without the fiber roots breaking. This will usually be from three to five days.

In washing we prefer to put it on the floor and turn a hose on it, and if you have a good pressure you will not need to touch the root with the hands. In any case do not scrub and scour the root. Just get the dirt off and stop. About one day after digging the root should be washed if a medium colored root is desired.

After your root is washed ready to dry there is still a half dozen ways of drying. Many prefer an upper room in the house for small lots. Spread the root on a table or bench about as high as the window stool. Then give it lots of air. Another good method is to subject it to a moderate artificial heat--from 60 to 90 degrees. We have seen some very nice samples of dry root where the drying was all done on the roof of some building, where it was exposed to the sun and dew, but was protected from rain. The slower the drying the darker the root.

Many suppose it is a difficult task to properly dry the Ginseng root, but it is not. The one essential is time. The operation cannot be fully and properly completed in much less than one month's time. Of course it should be dried fast enough so it will not sour, rot or mould. If you take a look at the root every day you can readily see if it is going too slow

CHAPTER IX. 64

and, if you find it is, at once use artificial heat for a few hours or days if need be. No diseased or unsound root should ever be dried. After the root is once dry it should be stored in dry place. Early fall generally is a poor time to sell as the Chinese exporters usually crowd the price down at that time.

In the Southern States artificial heat is seldom needed as the weather is usually warm enough to cure the roots about as they should be. In the Northern States, such as Minnesota, Wisconsin, Michigan, New York and New England States, cold and frosty nights and chilly days usually come in October, and sometimes in September, so that artificial heat is generally required to properly dry fall dug roots.

The statistics as published were compiled by Belt, Butler Co., buyers of Ginseng, 140 Greene St., New York:

Average prices for wild Ginseng, Sept. 1st, 1886, $1.90 Average prices for wild Ginseng, Sept. 1st, 1887, $2.10 Average prices for wild Ginseng, Sept. 1st, 1888, $2.30 Average prices for wild Ginseng, Sept. 1st, 1889, $2.85 Average prices for wild Ginseng, Sept. 1st, 1890, $3.40 Average prices for wild Ginseng, Sept. 1st, 1891, $3.40 Average prices for wild Ginseng, Sept. 1st, 1892, $3.00 Average prices for wild Ginseng, Sept. 1st, 1893, $3.00 Average prices for wild Ginseng, Sept. 1st, 1894, $3.50 Average prices for wild Ginseng, Sept. 1st, 1895, $3.25 Average prices for wild Ginseng, Sept. 1st, 1896, $4.10 Average prices for wild Ginseng, Sept. 1st, 1897, $3.25 Average prices for wild Ginseng, Sept. 1st, 1898, $4.00 Average prices for wild Ginseng, Sept. 1st, 1899, $6.00 Average prices for wild Ginseng, Sept. 1st, 1900, $5.00 Average prices for wild Ginseng, Sept. 1st, 1901, $5.50 Average prices for wild Ginseng, Sept. 1st, 1902, $5.10 Average prices for wild Ginseng, Sept. 1st, 1903, $6.20 Average prices for wild Ginseng, Sept. 1st, 1904, $7.40 Average prices for wild Ginseng, Sept. 1st, 1905, $7.00 Average prices for wild Ginseng, Sept. 1st, 1906, $7.00 Average prices for wild Ginseng, Sept. 1st, 1907, $7.00

The prices as published, it will be noticed, were average prices paid for wild Ginseng September 1 of each year. Wild Ginseng has usually sold higher in the season, say October and November. Late in the season of 1904 it sold for $8.50 for good Northern root, which we believe was

CHAPTER IX.

the top notch for average lots.

From 1860 to 1865, Ginseng ranged from 66c to 85c per lb., and from that period until 1899 it gradually increased in price until in September of that year it brought from $3.50 to $6.50 per lb., according to price and quality. In 1900 prices ruled from $3.00 to $5.75 per lb., but this was due to the war then existing in China which completely demoralized the market.

In 1901 prices ranged from $3.75 to $7.25 1902 prices ranged from 3.50 to 6.25 1903 prices ranged from 4.75 to 7.50 1904 prices ranged from 5.50 to 8.00 1905 prices ranged from 5.50 to 7.50 1906 prices ranged from 5.75 to 7.50 1907 prices ranged from 5.75 to 7.25

These prices cover the range from Southern to best Northern root.

The above information was furnished from the files of Samuel Wells & Co., Cincinnati, Ohio, the firm which has been in the "seng" business for more than half a century.

* * *

U. S. GOVERNMENT REPORTS.

Year Pounds Average price exported. per lb. 1858 366,055 $ .52 1868 370,066 1.02 1878 421,395 1.17 1888 308,365 2.13 1898 174,063 3.66 1901 149,069 5.30

* * *

Export of Ginseng for ten months ending April, 1908, was 144,533 pounds, valued at $1,049,736, against 92,650, valued at $634,523, for ten months ending April, 1907, and 151,188 pounds, valued at $1,106,544 for ten months ending April, 1906.

Since 1858 Ginseng has advanced from 52 cents a pound to $8.00 in 1907 for choice lots, an advance of 1400%.

CHAPTER IX.

In September, 1831, Ginseng was quoted to the collector at 15 to 16 cents per pound.

In the first place, practically all the Ginseng grown or collected from the woods in this country is exported, nearly all of it going to China, where it is used for medicinal purposes. The following figures are taken from the advanced sheets of the Monthly Summary of Commerce and Finance issued by the United States Department of Commerce and Labor. In the advanced sheets for June, 1906, we find under exports of Domestic Merchandise the following item:

Twelve Months Ending June.

Ginseng lbs. 1904 131,882 $851,820 1905 146,586 $1,069,849 1906 160,959 $1,175,844

From these figures it is clear that the Ginseng crop is of considerable proportions and steadily increasing. It is classed with chemicals, drugs, dyes and medicines and is in its class equaled or exceeded in value by only three things: copper sulphate, acetate of lime and patent medicines. These figures include, of course, both the wild and cultivated root. A little investigation, however, will soon convince any one that the genuine wild root has formed but a small portion of that exported in the last three years. This is for the very good reason that there is practically no wild root to be found. It has been all but exterminated by the "seng digger," who has carefully searched every wooded hillside and ravine to meet the demand of the last few years for green roots for planting. Practically all of the Ginseng now exported will of necessity be cultivated. Of all the Ginseng exported from this country, New York State very probably supplies the greater part. It was in that state that the cultivation of the plant originated and it is there that the culture has become most extensive and perfected. The largest garden in this country, so far as known, is that of the Consolidated Ginseng Company of New York State. Here about ten acres are under shade, all devoted to the growing of Ginseng. The crop is certainly a special one, to be successfully grown only by those who can bring to their work an abundance of time and intelligent effort. For those who are willing to run the risks of loss from diseases and who can afford to wait for returns on their investment, this crop offers relatively large profits.

[Illustration: A Three Year Old Cultivated Root.]

It is very simple to prepare a few wild roots for market. Wash them thoroughly, this I do with a tooth or nail brush, Writes a Northern grower, as they will remove the dirt from the creases without injury. Only a few roots should be put in the water at once as it does not benefit them to soak.

I have usually dried wild roots in the sun, which is the best way, but never put roots in the hot sun before the outside is dry, as they are apt to rot.

The cultivated root is more difficult to handle. They are cleaned the same as wild roots. On account of size and quality they have to be dried differently. My first cultivated roots were dried around the cook stove, which will answer for a few roots, providing the "lady of the house" is good natured.

Last year I dried about 500 pounds of green roots and so had to find something different. I made a drier similar to Mr. Stanton's plan, i. e., a box any size to suit the amount of roots you wish to dry. The one I made is about two feet by two and a half feet and two and one-half feet high, with one side open for the drawers to be taken out. The drawers are made with screen wire for bottom.

They should be at least two inches deep and two and one-half inches would be better. I bored a three-fourth-inch hole in the top a little ways from each corner and five in the center in about ten inches square, but now I have taken the top off, as I find they dry better.

I started this on the cook stove, but did not like it as I could not control the heat. As I had two Blue Flame oil stoves I tried it over one of them and it worked fine.

They were three-hole stoves, so I laid a board across each end for the drier to rest on. The drier has a large nail driven in each corner of the bottom so that it was four inches above the stove. Then I fixed a piece

CHAPTER IX. 68

of galvanized iron about 10x20 inches so that it was about two inches above top of stove, for the heat to strike against and not burn the roots.

At first I left out two of the lower drawers for fear of burning them. I only used the middle burner--and that turned quite low. I tried the flame with my hand between the stove and roots so as not to get it too high.

In this way I could get a slow heat and no danger of burning, which is the main trouble with drying by stove. It would take from two to four days to dry them, according to size. As soon as they were dried they were put in open boxes so if there was any moisture it could dry out and not mould, which they will do if closed up tight.

In using an oil stove one should be used that will not smoke. Never set the roots over when the stove is first lighted and they should be removed before turning the flame out, as they are apt to get smoked. Do not set stove in a draft.

In packing the dry root in boxes I break off the fine fiber, then they are ready for market.

Some time prior to 1907, or since cultivated Ginseng has been upon the market, its value has been from $1.00 to $2.00 per pound less than the wild and not in as active demand, even at that difference, as the wild. Today the value is much nearer equal. At first those engaged in the cultivation of Ginseng made the soil too rich by fertilizing and growth of the roots was so rapid that they did not contain the peculiar scent or odor of the genuine or wild. Of late years growers have learned to provide their plants with soil and surroundings as near like nature as possible. To this can largely be attributed the change.

Preparing the Roots for Market.

The roots are dug in the autumn, after the tops have died. Great care is taken not to bruise or injure them. They are then washed in rain water, the soil from all crevices and cracks being carefully cleaned away by a soft brush. Then they are wiped on a soft absorbent cloth, and are ready to be dried for market. The roots should never be split in washing or drying. It is of great importance, too, that the little neck or

CHAPTER IX. 69

bud-stem should be unbroken, for if missing the root loses two-thirds of its value in Chinese eyes. The roots may be dried in the sun or in a warm, dry room, but never over a stove or fire. Some growers have a special drier and use hot air very much on the principle of an evaporator. This does the work quickly and satisfactorily. As soon as the little fibrous roots are dry enough, they arc either clipped off or rubbed away by hand, and the root returned to the drier to be finished. The more quickly the roots are dried the better, if not too much heated. Much of the value of the product depends on the manner in which it is cured. This method is the one usually employed in America, but the Chinese prepare the root in various ways not as yet very well understood in the United States. Their preparation undoubtedly adds to the value of the product with the consumer.

Importance of Taste and Flavor.

Soils and fertilizers have a marked influence on products where taste and flavor is important, as with tobacco, coffee, tea, certain fruits, etc. This is true of Ginseng in a very marked degree. To preserve the flavor which marks the best grade of Ginseng, by which the Chinese judge it, it is essential that the soil in the beds should be as near like the original native forest as possible. Woods earth and leaf mold should be used in liberal quantities. Hardwood ashes and some little bone meal may be added, but other fertilizers are best avoided to be on the safe side.

When the chief facts of Ginseng culture had been ascertained, it naturally followed that some growers attempted to grow the biggest, heaviest roots possible in the shortest time, and hence fertilized their beds with strong, forcing manures, entirely overlooking the question of taste or flavor. When these roots were placed on the market the Chinese buyers promptly rejected them or took them at very low prices on account of defective quality. This question of flavor was a new problem to American buyers, for the reason stated and one which they were not prepared to meet at a moment's notice. Hence there has been a tendency with some exporters to be shy of all cultivated roots (fearing to get some of these "off quality" lots) until they were in position to test for flavor or taste by expert testers, as is done with wines, teas, coffees, tobaccos and other products where flavor is

CHAPTER IX. 70

essential.

This mistake led to the belief with some that the cultivated root is less valuable than the wild, but the very reverse is true. It has been proven by the fact that until these "off quality" lots appeared to disturb the market and shake confidence for the time being, cultivated roots have always commanded a much better price per pound than uncultivated. The grower who freely uses soil from the forest and lets forcing fertilizers severely alone, has nothing to fear from defective quality, and will always command a good price for his product.

Ginseng should only be dug for the market late in the fall. In the spring and summer the plant is growing and the root is taxed to supply the required nutriment. After the plant stops growing for the season the root becomes firm and will not dry out as much as earlier in the season. It takes four to five pounds of the green root early in the season to make one of dry; later three green will make one of dry.

In the Ginseng, like many other trades, there are tricks. In some sections they practice hollowing out roots while green and filling the cavity with lead or iron. When Ginseng is worth four or five dollars per pound and lead or iron only a few cents, the profit from this nefarious business can be seen. The buyers have "got on to" the practice, however, and any large roots that appear too heavy are examined. The filling of roots with lead, etc., has about had its day.

Seng should be dug and washed clean before it wilts or shrinks; it should then be dried in the shade where the dust and dirt cannot reach it and should not be strung on strings. The roots should be handled carefully so as not to break them up, the more fiber the less the value, as well as size which helps to determine the value.

The collecting of the root for the market by the local dealer has its charm; at least one would think so, to see how eagerly it is sought after by the collector, who often finds when he has enough for a shipment that he faces a loss instead of a profit. The continual decrease in the annual output of the root should produce a steadily advancing market. The price does advance from year to year, but the variation in the price of silver and the scheming of the Chinamen produces crazy spurts in the price of the root.

Present prices are rather above average, but little can be predicted about future conditions. Chinese conservatism, however, leads us to believe present prices will continue.

CHAPTER X.

LETTERS FROM GROWERS.

The culture of Ginseng has a pioneer or two located in this part of the country (N. Ohio), and having one-fourth of an acre under cultivation myself, it was with interest that I visited some of these growers and the fabulous reports we have been reading have not been much exaggerated, in my estimation, but let me say right here they are not succeeding with their acres as they did with their little patch in the garden. One party gathered 25 pounds of seed from a bed 40x50 feet last season, and has contracted 30 pounds of the seed at $36 per pound, which he intends to gather from this bed this season. He then intends to dig it, and I will try to get the facts for this magazine.

Now, to my own experience. I planted three hundred roots in the fall of '99. The following season from the lack of sufficient shade they failed to produce any seed; I should have had two or three thousand seed. Understand, these were wild roots just as they were gathered from the forest.

In 1901 I gathered about one pound or 8,000 seed, in 1902 three pounds and am expecting 30,000 seed from these 300 plants this season. Last season I gathered 160 seed from one of these plants and 200 seed bunches are not uncommon for cultivated roots to produce at their best. I have dug no roots for market yet, as there has been too great a demand for the seed. My one-fourth acre was mostly planted last season, and is looking very favorable at the present time. It is planted in beds 130 feet long and 5 feet wide; the beds are ridged up with a path and ditch 2 feet across from plant to plant, making the beds, including the paths, 7 feet wide. Beds arranged in this manner with the posts that support the shade set in the middle of the beds are very convenient to work in, as you do not have to walk in the beds, all the work being done from the paths.

My soil is a clay loam and it was necessary for me to place a row of tile directly under one bed; this bed contains 1,000 plants and has been planted two years, and I find the tile a protection against either dry or wet weather; I shall treat all beds in a like manner hereafter.

CHAPTER X.

If you are thinking of going into the Ginseng business and your soil is sand or gravel, your chances for success are good; if your soil is clay, build your beds near large trees on dry ground or tile them and you will come out all right. In regard to the over-production of this article, would say that dry Ginseng root is not perishable, it will keep indefinitely and the producers of this article will not be liable to furnish it to the Chinaman only as he wants it at a fair market price.

W. C. Sorter, Lake County, Ohio.

* * *

Even in this thickly settled country, I have been able to make more money digging Ginseng than by trapping, and I believe that most trappers could do the same if they became experts at detecting the wild plant in its native haunts.

I have enjoyed hunting and trapping ever since I could carry a firearm with any degree of safety to myself, and have tramped thru woods full of Ginseng and Golden Seal for twenty years, without knowing it. Three years ago last summer I saw an advertisement concerning Ginseng Culture. I sent and got the literature on the subject and studied up all I could. Then I visited a garden where a few cultivated plants were grown, and so learned to know the plant. I had been told that it grew in the heavy timber lands along Rock River, so I thought I would start a small garden of some 100 or 200 roots.

The first half day I found 6 plants, and no doubt tramped on twice that many, for I afterward found them thick where I had hunted. The next time I got 26 roots; then 80, so I became more adept in "spotting" the plants, the size of my "bag" grew until in September I got 343 roots in one day. That fall, 1904, I gathered 5,500 roots and 2,000 or 3,000 seed. These roots and seed I set out in the garden in beds 5 feet wide and 40 feet long, putting the roots in 3 or 5 inches apart anyway, and the seeds broadcast and in rows. I mulched with chip manure, leaf mold and horse manure. Covered with leaves in the fall, and built my fence.

CHAPTER X.   74

The next spring the plants were uncovered and they came well. I believe nearly every one came up. They were too thick, but I left them. The mice had run all thru the seed bed and no doubt eaten a lot of the seed. That spring I bought 5,000 seed of a "seng" digger and got "soaked." The fall of 1905 I dug 500 more roots and harvested 15,000 seeds from my beds. The roots were planted in an addition and seed put down cellar. Last fall I gathered 5,500 more roots from the woods, grew about 3,000 seedlings in the garden and harvested 75,000 seeds. I dug a few of the older roots and sold them.

The worst enemy I find to Ginseng culture is Alternaria, of a form of fungus growth on the leaf of the plant. This disease started in my beds last year, but I sprayed with Bordeaux Mixture and checked it. I have not as yet been troubled with "damping off" of seedlings. I shall try Bordeaux if it occurs.

My garden is now 100 feet by 50 feet, on both sides of a row of apple trees, in good rich ground which had once been a berry patch. I used any old boards I could get for the side fence, not making it too tight. For shade I have tried everything I could think of. I used burlap tacked on frames, but it rotted in one season. I used willow and pine brush and throwed corn stalks and sedge grass on them. For all I could see, the plants grew as well under such shade as under lath, although the appearance of the yard is not so good. I also ran wild cucumbers over the brush and like them very well. They run about 15 feet, so they do not reach the center of the garden until late in the season. I planted them only around the edge of the garden.

[Illustration: Bed of Mature Ginseng Plants Under Lattice.]

In preparing my soil, I mixed some sand with the garden soil to make it lighter; also, woods earth, leaf mold, chip manure and barnyard manure, leaving it mostly on top. I take down the shade each fall and cover beds with leaves and brush. This industry is not the gold mine it was cracked up to be. The price is going down, lumber for yard and shade is going up. The older the garden, the more one has to guard against diseases, so one may not expect more than average returns for his time and work. Still I enjoy the culture, and the work is not so hard, and it is very interesting to see this shy wild plant growing in its

CHAPTER X.

new home.

In order to keep up the demand for Ginseng, we must furnish the quality the Chinese desire, and to do this, I believe we must get back to the woods and rotten oak and maple wood, leaf mold and the humid atmosphere of the deep woodlands. I have learned much during the short time I have been growing the plant, but have only given a few general statements.

John Hooper, Jefferson Co., Wis.

* * *

I believe most any one that lives where Ginseng will grow could make up a small bed or two in their garden and by planting large roots and shading it properly, could make it a nice picture. Then if they could sell their seed at a good price might make it profitable, but when it comes down to growing Ginseng for market I believe the only place that one could make a success would be in the forest or in new ground that still has woods earth in it and then have it properly shaded.

The finest garden I ever saw is shaded with strips split from chestnut cuts or logs. There are thousands of young "seng" in this garden from seedlings up to four years old this fall, and several beds of roots all sizes that were dug from the woods wild and are used as seeders. These plants have a spreading habit and have a dark green healthy look that won't rub off. It is enough to give "seng" diggers fits to see them.

I have my Ginseng garden in a grove handy to the house, where it does fairly well, only it gets a little too much sun. I have a few hundred in the forest, where it gets sufficient shade and there is a vast difference in the color and thriftiness of the two.

The seed crop will be a little short this fall in this section, owing to heavy frosts in May which blighted the blossom buds on the first seng that came up. My seed crop last fall was ten quarts of berries which are buried now in sand boxes. My plan for planting them this fall is to stick the seeds in beds about 4x4 inches.

CHAPTER X.

I see where some few think that the mulch should be taken off in the spring, which I think is all wrong. I have been experimenting for seven years with Ginseng and am convinced that the right way is to keep it mulched with leaves. The leaves keep the ground cool, moist and mellow and the weeds are not half so hard to keep down. It is surely the natural way to raise Ginseng.

My worst trouble in raising Ginseng is the damping off of the seedlings. My worst pest is chickweed, which grows under the mulch and seems to grow all winter. It seeds early and is brittle and hard to get the roots when pulling. Plantain is bad sometimes, the roots go to the bottom of the bed. Gladd weed is also troublesome. I think one should be very careful when they gather the mulch for it is an easy matter to gather up a lot of bad weed seed.

I see in the H-T-T where some use chip manure on their "seng" beds. I tried that myself, but will not use it again on seed beds any way. I found it full of slugs and worms which preyed on the seedlings. Sometimes cut worms cut off a good many for me. Grub worms eat a root now and then. Leaf rollers are bad some years, but the worst enemy of all is wood mice. If one does not watch carefully they will destroy hundreds of seed in a few nights.

I find the best way to destroy them is to set little spring traps where they can run over them. There was a new pest in this locality this year which destroyed a big lot of seed. It was a green cricket something like a katydid. They were hard to catch, too.

Thos. G. Fulcomer, Indiana Co., Pa.

[Illustration: Some Thrifty Plants--An Ohio Garden.]

The notions of the Chinese seem as difficult to change as the law of the Medes and Persians, and his notion that the cultivated article is no good, if once established, will always be established. This will be a sad predicament for the thousands who may be duped by the reckless Ginseng promoter. One principle of success in my business is to please the purchaser or consumer. This is the biggest factor in Ginseng culture.

CHAPTER X. 77

The Chinaman wants a certain quality of flavor, shape, color, etc., in his Ginseng, and as soon as the cultivators learn and observe his wishes so soon will they be on the right road to success. Ginseng has been brought under cultivation and by doing this it has been removed from its natural environments and subjected to new conditions, which are making a change in the root. The object of the Ginseng has been lost sight of and the only principle really observed has been to grow the root, disregarding entirely the notions of the consumer.

Thousands have been induced by the flattering advertisements to invest their money and begin the culture of Ginseng. Not one-half of these people are familiar with the plant in its wild state and have any idea of its natural environments. They are absolutely unfit to grow and prepare Ginseng for the Chinese market. Thousands of roots have been spoiled in the growing or in the drying by this class of Ginseng growers. Many roots have been scorched with too much heat, many soured with not the right conditions of heat, many more have been spoiled in flavor by growing in manured beds and from certain fertilizers. All these damaged roots have gone to the Chinese as cultivated root and who could blame him for refusing to buy and look superstitious at such roots?

Now as to profits. Not one-half the profits have been made as represented. Not one-half of those growing Ginseng make as much as many thousands of experienced gardeners and florists are making with no more money invested and little if any more labor and no one thinks or says anything about it. Many articles have appeared in the journals of the past few years, and when you read one you will have to read all, for in most part they have been from the over-stimulated mind of parties seeking to get sales for so-called nursery stock.

Probably the first man to successfully cultivate Ginseng was Mr. Stanton, of New York State. His gardens were in the forest, from this success many followed. Then the seed venders and wide publicity and the garden cultivation under lattice shade. Then the refusal of the Chinese to buy these inferior roots.

Now, it is my opinion the growers must return to the forest and spare no labor to see that the roots placed on the market are in accordance

CHAPTER X. 78

with the particular notions of the consumer. Ginseng growers may then hope to establish a better price and ready market for their root.

The color required by the Chinese, so far as my experiments go, come from certain qualities of soil. The yellow color in demand comes to those roots growing in soil rich in iron. The particular aromatic flavor comes from those growing in clay loam and abundant leaf mold of the forest. I have found that by putting sulphate of iron sparingly in beds and the roots growing about two years in this take on the yellow color.

I have three gardens used for my experiments, two in forest and one in garden. They contain altogether about twenty-five thousand plants. One garden is on a steep north hillside, heavily shaded by timber. These plants have a yellowish color and good aromatic taste. They have grown very slow here; about as much in three years as they grow in one year in the garden. The other forest garden is in an upland grove with moderate drain, clay loam and plenty of leaf mold; the trees are thin and trimmed high. The beds are well made, the roots are light yellow and good flavor, they grow large and thrifty like the very best of wild.

I am now planting the seed six inches apart and intend to leave them in the bed without molesting until matured. The beds under the lattice in the garden have grown large, thick, white and brittle, having many rootlets branching from the ends of the roots, The soil is of a black, sandy loam. They do not have the fine aromatic flavor of those roots growing in the woods.

The plants I have used in the most part were produced from the forest here in Minnesota and purchased from some diggers in Wisconsin. I have a few I procured from parties advertising seed and plants, but find that the wild roots and seeds are just as good for the purpose of setting if due care is exercised in sorting the roots.

There has been considerable said in the past season by those desiring to sell nursery stock condemning the commission houses and ignoring or minimizing the seriousness of the condition which confronts the Ginseng grower in a market for his root. Now, I believe the commission men are desirous of aiding the Ginseng growers in a market for his

CHAPTER X.

roots so long as the grower is careful in his efforts to produce an article in demand by the consumer.

In my opinion those who are desirous of entering an industry of this kind will realize the most profits in the long run if they devote attention to the study and cultivation of those medical plants used in the therapy of the regular practice of medicine, such as Hydrastis, Seneca, Sanguinaria, Lady Slipper, Mandrake, etc. They are easily raised and have a ready market at any of our drug mills. I have experimented with a number of these and find they thrive under the care of cultivation and I believe in some instances the real medical properties are improved, as Atropine in Belladonna and Hydrastine in Hydrastis.

I have several thousand Hydrastis plants under cultivation and intend to make tests this season for the quantity of Hydrastine in a given weight of Hydrastis and compare with the wild article. It is the amount of Hydrastine or alkaloid in a fluid extract which by test determines the standard of the official preparation and is the real valuable part of the root.

This drug has grown wonderfully in favor with the profession in recent years and this increased demand with decrease of supply has sent the price of the article soaring so that we are paying five times as much for the drug in stock today as we paid only three or four years ago.

I trust that I have enlarged upon and presented some facts which may be of interest and cause those readers who are interested in this industry to have a serious regard for the betterment of present conditions, to use more caution in supplying the market and not allow venders to seriously damage the industry by their pipe dream in an attempt to find sales for so-called nursery stock.

L. C Ingram, M. D., Wabasha County, Minn.

* * *

It was in the year of 1901, in the month of June, that I first heard of the wonderful Ginseng plant. Being a lover of nature and given to strolling in the forests at various times, I soon came to know the Ginseng plant in its wild state.

Having next obtained some knowledge regarding the cultivation of this plant from a grower several miles away, I set my first roots to the number of 100 in rich, well-drained garden soil, over which I erected a frame and covered it with brush to serve as shade.

In the spring of 1902 nearly all the roots made their appearance and from these I gathered a nice crop of seed later on in the season. That summer I set out 2,200 more wild roots in common garden soil using lath nailed to frames of scantling for shade. Lath was nailed so as to make two-thirds of shade to one-third of sun. This kind of shading I have since adopted for general use, because I find it the most economical and for enduring all kinds of weather it cannot be surpassed.

During the season of 1903 I lost several hundred roots by rot, caused by an excessive wet season and imperfect drainage.

In the seasons of 1903 and 1904 I set about 2,000 wild roots in common garden soil, mixed with sand and woods dirt and at this writing (July 9th, 1905) some of these plants stand two feet high, with four and five prongs on branches, thus showing the superiority of this soil over the others I have previously tried.

[Illustration: New York Grower's Garden.]

During my five years of practical experience in the cultivation of this plant I have learned the importance of well drained ground, with porous open sub-soil for the cultivation of Ginseng. My experience with clay hard-pan with improper drainage has been very unsatisfactory, resulting from the loss of roots by rot. Clay hard-pan sub-soil should be tile-drained.

Experience and observation have taught me that Ginseng seed is delicate stuff to handle and it is a hard matter to impress upon people the importance of taking care of it. I have always distinctly stated that it must not be allowed to get dry and must be kept in condition to promote germination from the time it is gathered until sown. Where a

CHAPTER X.

consider able quantity is to be cared for, the berries should be packed in fine, dry sifted sand soon after they are gathered, using three quarts of sand and two quarts of berries. The moisture of the berries will dampen the sand sufficiently. But if only a few are to be packed the sand should be damp.

Place one-half inch sand in box and press smooth. On this place a layer of berries; cover with sand, press, and repeat the operation until box is full, leaving one-half inch of sand on top; on this place wet cloth and cover with board. Place box in cellar or cool shady place. The bottom of the box should not be tight. A few gimlet holes with paper over them to keep the sand from sifting thru will be all right. Any time after two or three months, during which time the seeds have lost their pulp and nothing but the seed itself remains, seed may be sifted out, washed, tested and repacked in damp sand until ready to sow.

Best Time to Sow Seed.

Since it takes the seed eighteen months to germinate, seed that has been kept over one season should be planted in August or September. I like to get my old crop of seed out of the way before the new crop is harvested, and also because my experience has been that early sowing gives better results than late.

One should be careful in building his Ginseng garden that he does not get sides closed too tight and thus prevent a free circulation of air going thru the garden, for if such is the case during a rainy period the garden is liable to become infected with the leaf spot and fungus diseases.

The drop in price of cultivated root was caused chiefly thru high manuring, hasty and improper drying of the root. In order to bring back the cultivated root to its former standing among the Chinese, we must cease high manuring and take more pains and time in drying the root, and then we will have a steady market for American cultivated root for years to come.

J. V. Hardacre, Geauga County, Ohio.

CHAPTER X.

* * *

In 1900 I went to the woods and secured about fifty plants of various sizes and set them in the shade of some peach and plum trees in a very fertile spot. They came up in 1901, that is, part of them did, but the chickens had access to them and soon destroyed the most of them, that is, the tops.

In 1902 only a few bunches came up, and through neglect (for I never gave them any care) the weeds choked them and they did no good. In 1903 the spirit of Ginseng growing was revived in me and I prepared suitable beds, shade and soil, and went to work in earnest. I secured several more plants and reset those that I had been trying to grow without care. In 1904 my plants came up nicely. I also secured several hundred more plants and set them in my garden.

The plants grew well and I harvested about 1,000 seed in the fall. Several Ginseng gardens were injured by a disease that seemed to scald the leaves and then the stalk became affected. In a short time the whole top of the plant died, but the root remained alive. My Ginseng was not affected in this way, or at least I did not notice it.

In 1905 I had a nice lot of plants appear and they grew nicely for a while, and as I was showing a neighbor thru the garden he pointed out the appearance of the disease that had affected most of the gardens in this county the previous year, and was killing the tops off of all the Ginseng in them this year. I began at once to fight for the lives of my plants by cutting off all affected parts and burning them.

I also took a watering pot and sprinkled the plants with Bordeaux Mixture. This seemed to help, and but few of the plants died outright.

I harvested several thousand seed. I placed the seed in a box of moist sand and placed them in the cellar and about one-third of them were germinated by the following spring, and there was not another garden in this vicinity, to my knowledge, that secured any seed. This fact caused me to think that spraying with Bordeaux Mixture would check the disease. It was certain that if the disease could not be prevented or quit of its own accord, Ginseng could not be grown in this county.

CHAPTER X. 83

In 1906 my plants came up nicely and grew as in the previous season. I noticed the disease on some of the plants about the last of May so I began removing the affected parts, also to sprinkle with Bordeaux Mixture with about the same results as the year before. In the fall I harvested about twelve or fifteen thousand seed.

I might say here that I sprinkled the plants about every two or three weeks. I raised the only seed that was harvested in this vicinity, and most all the large "seng" was dried and sold out of their gardens.

Early in 1907 I secured a compressed air sprayer, for I had come to the conclusion that spraying would be lots better than sprinkling. On the appearance of the first plants in the spring I began spraying and sprayed every week or ten days until about the first of September. I saved the life of most of my plants.

For an experiment I left about five feet of one bed of two-year-old plants unsprayed. It grew nicely until about the 10th of June, then the disease struck it, and in about two or three weeks it was about all dead, while the remainder that was sprayed lived thru till frost, and many of them bore seed. I harvested about 20,000 seed in the fall.

I believe if I had not persisted in the spraying I would not have harvested one fully matured seed, for none of my neighbors secured any. In September, 1906, I dug one bed of large roots thinly set on a bed 4x16 feet which netted me $8.49.

In September, 1907, I dug a bed 4x20 feet which netted me $19.31.

This is my experience. Of course I have omitted method of preparing beds, shade, etc.

A. C Herrin, Pulaski County, Ky.

* * *

Many inquiries are continually being received concerning Ginseng, Some of the many questions propounded are as follows: Is Ginseng growing profitable? Is it a difficult crop to grow? How many years will it

CHAPTER X.

take to grow marketable roots? When is the best time to set plants and sow the seed? What kind of soil is best adapted to the crop? Does the crop need shade while growing? Do the tops of Ginseng plants die annually? Must the roots be dried before marketable? What time of year do you dig the roots? Does the cultivation of the plants require much labor? What are the roots used for and where does one find the best markets? About what are the dry roots worth per pound? How are the roots dried? How many roots does it take to make a pound? Have you sold any dry roots yet from your garden? How long does it take the seed of Ginseng to germinate?

Do you sow the seeds broadcast or plant in drills? How far apart should the plants be set? Do you mulch beds in winter? Is it best to reset seedlings the first year? How many plants does it require to set an acre? What is generally used for shading? Has the plant or root any enemies? When does the seed ripen? How wide do you make your beds? Do you fertilize your soil? Will the plants bear seed the first year? What price do plants and seed usually bring? What does the seed look like?

It will be almost impossible to answer all of the above questions, but will try to give a few points regarding Ginseng and Ginseng growing which may help some reader out. In the spring of 1899 I began experimenting with a few Ginseng plants, writes an Indiana party, and at present have thousands of plants coming along nicely from one to seven years old. Last fall I planted about eight pounds of new seed. The mature roots are very profitable at present prices. They are easily grown if one knows how. It takes about five years to grow marketable roots.

The seed is planted in August and September; the plants set in September and October. A rich, dark sandy loam is the most desirable soil for the crop, which requires shade during growth. The plants are perennial, dying down in the fall and reappearing in the spring. The roots must be dried for market. They should be dug some time in October. Cultivation of the crop is comparatively simple and easy. The crop is practically exported from this country to China, where the roots are largely used for medicinal purposes. The best prices are paid in New York, Chicago, Cincinnati and San Francisco. Dry roots usually

CHAPTER X. 85

bring $4.00 to $8.00 per pound as to quality. The drying is accomplished the same way fruit is dried. The number of roots in a pound depends on their age and size.

The seed of Ginseng germinates in eighteen months. Sow the seed in drill rows and set the plants about eight inches apart each way. Mulch the beds with forest leaves in the fall. The seedlings should be reset the first year. It requires about 100,000 plants to cover an acre. The shade for the crop is usually furnished by the use of lath or brush on a stationary frame built over the garden.

Moles and mice are the only enemies of Ginseng and sometimes trouble the roots, but are usually quite easily kept out. The seed of Ginseng ripens in August. Seed beds are usually made four feet wide. The best fertilizer is leaf mould from the woods. The plants will not bear much seed the first year. The price of both seed and plants varies considerably. The seed looks like those of tomatoes, but is about ten times larger.

Ginseng is usually found growing wild in the woods where beech, sugar and poplar grow. The illustration shows a plant with seed. Early in the season, say June and early July, there is no stem showing seed. (See cover.)

The plant usually has three prongs with three large leaves and has small ones on each stem. Note the illustration closely. Sometimes there are four prongs, but the number of leaves on each prong is always five--three large and two small.

The leading Ginseng states are West Virginia, Kentucky and Tennessee. It is also found in considerable quantities in Virginia, Pennsylvania, Ohio, Indiana, Illinois, Michigan, New York, and even north into Southern Canada. It is also found in other Central and Southern states.

During the past few years the wild root has been dug very close, and in states where two or three years ago Ginseng was fairly plentiful is now considerably thinned out. In some sections "sengers" follow the business of digging the wild root from June to October. They make good wages quite often. It is these "sengers" that have destroyed the wild crop and paved the way for the growers. The supply of wild root will no doubt become less each year, unless prices go down so that there will not be the profit in searching for it.

CHAPTER XI.

GENERAL INFORMATION.

Cultivated root being larger than wild takes more care in drying. Improper drying will materially impair the root and lessen its value.

It is those who study the soil and give attention to their fruit that make a success of it. The same applies to growing Ginseng and other medicinal plants.

When buying plants or seeds to start a garden it will be well to purchase from some one in about your latitude as those grown hundreds of miles north or south are not apt to do so well.

Ginseng culture is now carried on in nearly all states east of the Mississippi River as well as a few west. The leading Ginseng growing states, however, are New York, Michigan, Ohio, Kentucky and Minnesota.

Thruout the "Ginseng producing section" the plants are dug by "sengers" from early spring until late fall. The roots are sold to the country merchants for cash or exchanged for merchandise. The professional digger usually keeps his "seng" until several pounds are collected, when it is either shipped to some dealer or taken to the county seat or some town where druggists and others make the buying of roots part of their business. Here the digger could always get cash for roots which was not always the case at the country store.

Quite often we hear some one say that the Chinese will one of these days quit using Ginseng and there will be no market for it. There is no danger, or at least no more than of our people giving up the use of tea and coffee. Ginseng has been in constant use in China for hundreds of years and they are not apt to forsake it now.

The majority of exporters of Ginseng to China are Chinamen who are located in New York and one or two cities on the Pacific coast. There is a prejudice in China against foreigners so that the Chinamen have an advantage in exporting. Few dealers in New York or elsewhere

CHAPTER XI. 88

export--they sell to the Chinamen who export.

The making of Bordeaux Mixture is not difficult. Put 8 pounds bluestone in an old sack or basket and suspend it in a 50-gallon barrel of water. In another barrel of same size, slack 8 pounds of good stone lime and fill with water. This solution will keep. When ready to use, stir briskly and take a pail full from each barrel and pour them at the same time into a third barrel or tub. This is "Bordeaux Mixture." If insects are to be destroyed at the same time, add about 4 ounces of paris green to each 50 gallons of Bordeaux. Keep the Bordeaux well stirred and put on with a good spray pump. Half the value in spraying is in doing it thoroughly.

It is our opinion that there will be a demand for Seneca and Ginseng for years. The main thing for growers to keep in mind is that it is the wild or natural flavor that is wanted. To attain this see that the roots are treated similar to those growing wild. To do this, prepare beds of soil from the woods where the plants grow, make shade about as the trees in the forests shade the plants, and in the fall see that the beds are covered with leaves. Study the nature of the plant as it grows wild in the forest and make your "cultivated" plants "wild" by giving them the same conditions as if they were growing wild in the forest. As mentioned in a former number, an easy way to grow roots is in the native forest. The one drawback is from thieves.

The above appeared as an editorial in the Hunter-Trader-Trapper, August, 1905.

Growing Ginseng and Golden Seal will eventually become quite an industry, but as we have said before, those that make the greatest success at the business, will follow as closely as possible the conditions under which the plants grow in the forests, in their wild state. Therein the secret lies. There is no class of people better fitted to make a success at the business than hunters and trappers, for they know something of its habits, especially those of the Eastern, Central and Southern States, where the plants grow wild. There is no better or cheaper way to engage in the business than to start your "garden" in a forest where the plant has grown. Forests where beech, sugar and poplar grow are usually good for Ginseng. The natural forest shade is

CHAPTER XI.

better than the artificial.

[Illustration: Forest Bed of Young "Seng." These Plants, However, Are too Thick.]

This is a business that hunters and trappers can carry on to advantage for the work on the "gardens" is principally done during the "off" hunting and trapping season.

The writer has repeatedly cautioned those entering the business of Ginseng culture to be careful. The growing of Ginseng has not proven the "gold mine" that some advertisers tried to make the public believe, but at the same time those who went at the business in a business-like manner have accomplished good results--have been well paid for their time. In this connection notice that those that have dug wild root for years are the most successful. Why? Because they are the ones whose "gardens" are generally in the forests or at least their plants are growing under conditions similar to their wild state. Therein the secret lies.

The majority of farmers, gardeners, etc., know that splendid sweet potatoes are grown in the lands of the New Jersey meadows. The potatoes are known thruout many states as "Jersey Sweets" and have a ready sale. Suppose the same potato was grown in some swampy middle state, would the same splendid "Jersey Sweet" be the result? Most assuredly not. If the same kind of sandy soil which the sweet potato thrives in in New Jersey is found the results will be nearer like the Jersey.

Again we say to the would-be grower of medicinal roots or plants to observe closely the conditions under which the roots thrive in their wild state and cultivate likewise, that is, grow in the same kind of soil, same density of shade, same kind and amount of mulch (leaves, etc.) as you observe the wild plant.

The growing of medicinal plants may never be a successful industry for the large land owner, for they are not apt to pay so much attention to the plants as the person who owns a small place and is engaged in fruit growing or poultry raising. The business is not one where acres

CHAPTER XI. 90

should be grown, in fact we doubt if any one will ever be successful in growing large areas. The person who has acres of forest land should be able to make a good income by simply starting his "gardens in the woods." The shade is there, as well as proper mulch, etc. In fact it is the forest where most of the valuable medicinal plants grow of their own accord. The conditions of the soil are there to produce the correct flavor. Some of the growers who are trying to produce large roots quickly are having trouble in selling their production. The dealers telling them that their roots have not the wild natural flavor--but have indications of growing too quickly and are probably cultivated.

While plants can be successfully cultivated by growing under conditions similar to the forest yet if there are forest lands near, you had better make your "gardens" there. This will save shading. In the north, say Canada, New England and states bordering on Canada, shading need not be so thick as farther south. In those states, if on high land, even a south slope may be used.

In other states a northern or eastern slope is preferred, altho if the shading is sufficiently heavy "gardens" thrive. Read what the various growers say before you start in the business, for therein you will find much of value. They have made mistakes and point these out to others.

From 1892 to 1897 the writer was on the road for a Zanesville, Ohio, firm as buyer of raw furs, hides, pelts and tallow. The territory covered was Southern Ohio, Western Pennsylvania, West Virginia and Northern Kentucky. During that time Ginseng was much more plentiful than now. Once at Portsmouth a dealer from whom I occasionally bought hides, had 21 sugar barrels full of dried seng--well on to 3,000 pounds. It was no uncommon thing to see lots of 100 to 500 pounds. I did not make a business of buying seng and other roots, as it was not handled to any great extent by the house I traveled for, altho I did buy a few lots ranging from 5 to 100 pounds, The five years that I traveled the territory named I should say that I called upon dealers who handled 100,000 pounds or 20,000 annually. This represented probably one-fifth of the collection. These dealers of course had men out.

CHAPTER XI.

Just what the collection of Ginseng in that territory is now I am unable to say as I have not traveled the territory since 1900, but from what the dealers and others say am inclined to think the collection is only about 10% what it was in the early '90s.

This shows to what a remarkable extent the wild root has decreased. The same decrease may not hold good in all sections, yet it has been heavy and unless some method is devised the wild root will soon be a thing of the past.

Diggers should spare the young plants. These have small roots and do not add much in value to their collection. If the young plants were passed by for a few years the production of the forest--the wild plant--could be prolonged indefinitely.

A root buyer for a Charleston, W. Va., firm, who has traveled a great deal thru the wild Ginseng sections of West Virginia, Kentucky, Tennessee, Indiana and Ohio says: The root is secured in greatest quantities from the states in the order named. Golden seal is probably secured in greatest quantities from the states as follows: West Virginia, Kentucky, Ohio, Missouri, Pennsylvania. A great deal is also secured from Western States and the North.

The "sengers" start out about the middle of May, altho the root is not at the best until August. At that time the bur is red and the greatest strength is in the root.

Many make it a business to dig seng during the summer. Some years ago I saw one party of campers where the women (the entire family was along) had simply cut holes thru calico for dresses, slipping same over the head and tied around the waist--not a needle or stitch of thread had been used in making these garments.

Some of these "sengers" travel with horses and covered rig. These dig most of the marketable roots. Others travel by foot carrying a bag to put Ginseng in over one shoulder and over the other a bag in which they have a piece of bacon and a few pounds of flour. Thus equipped they stay out several days. The reason these men only dig Ginseng is that the other roots are not so valuable and too heavy to carry.

Sometimes these men dig Golden Seal when near the market or about ready to return for more supplies.

Some years ago good wages were made at digging wild roots but for the past few years digging has been so persistent that when a digger makes from $1.00 to $2.00 per day he thinks it is good.

Some say that the Ginseng growing business will soon be overdone and the market over-supplied and prices will go to $1.00 per pound or less for dried root. If all who engage in the business were able to successfully grow the plant such might be the case. Note the many that have failed. Several complain that their beds in the forests are infested with many ups and downs from such causes as damp blight, root rot, animals and insect pests. A few growers report that mice did considerable damage in the older beds by eating the neck and buds from the roots.

There seems to be a mistaken idea in regard to "gardens in the forest." Many prepare their beds in the forests, plant and cultivate much the same as the grower under artificial shade. While this is an improvement over the artificial shade, fertilized and thickly planted bed, it is not the way that will bring best and lasting results.

Why? Because plants crowded together will contract diseases much sooner than when scattered. One reason of many failures is that the plants were too thick. Those that can "grow" in the forests are going to be the ones that make the greatest success. Farmers, horticulturists, gardeners, trappers, hunters, guides, fishermen who have access to forest land should carefully investigate the possibilities of medicinal root culture.

Those who have read of the fortune to be made at growing Ginseng and other medicinal roots in their backyard on a small plat (say a rod or two) had best not swallow the bait. Such statements were probably written by ignorant growers who knew no better or possibly they had seed and plants for sale. Ginseng growing, at best, should be done by persons who know something of plants, their habits, etc, as well as being familiar with soil and the preparation of same for growing crops.

CHAPTER XII.

MEDICINAL QUALITIES.

In reply to E. T. Flanegan and others who wish to know how to use Ginseng as a medicine, I will suggest this way for a general home made use, says a writer in Special Crops. Take very dry root, break it up with a hammer and grind it thru a coffee mill three or four times till reduced to a fine powder. Then take three ounces of powder and one ounce of milk sugar. To the milk sugar add sixty drops of oil of wintergreen and mix all the powders by rubbing them together and bottle. Dose one teaspoonful, put into a small teacupful of boiling water. Let it stay a little short of boiling point ten minutes. Then cool and drink it all, hot as can be borne, before each meal. It may be filtered and the tea served with cream and sugar with the meal. Made as directed this is a high grade and a most pleasant aromatic tea and has a good effect on the stomach, brain and nervous system. To those who have chronic constipation, I would advise one fourth grain of aloin, taken every night, or just enough to control the constipation, while taking the Ginseng tea. If the evening dose of Ginseng be much larger it is a good safe hypnotic, producing good natural sleep.

The writer prefers the above treatment to all the whiskey and patent medicine made. To those who are damaged or made nervous by drinking coffee or tea, quit the coffee or tea and take Ginseng tea as above directed. It is most pleasant tasted and a good medicine for your stomach. I do not know just how the Chinese prepare it into medicine, but I suppose much of it is used in a tea form as well as a tincture. As it is so valuable a medicine their mode of administration has been kept a secret for thousands of years. There must be some medical value about it of great power or the Chinese could not pay the price for it. It has been thought heretofore that the Chinese were a superstitious people and Used Ginseng thru ignorance, but as we get more light on the medical value of the plant the plainer it gets that it is us fellows--the Americans--that have been and are yet in the "shade" and in a dark shade, too. We think the time not far off when it will be recognized as a medical plant and a good one, too, and its great medical value be made known to the world.

CHAPTER XII.

For several years past I have been experimenting with Ginseng as a medical agent and of late I have prescribed, or rather added it, to the treatment of some cases of rheumatism. I remember one instance in particular of a middle-aged man who had gone the rounds of the neighborhood doctors and failed of relief, when he employed me. After treating him for several weeks and failing to entirely relieve him, more especially the distress in bowels and back, I concluded to add Ginseng to his treatment. After using the medicine he returned, saying the last bottle had served him so well that he wanted it filled with the same medicine as before. I attribute the curative properties of Ginseng in rheumatism to stimulating to healthy action of the gastric juices; causing a healthy flow of the digestive fluids of the stomach, thereby neutralizing the extra secretion of acid that is carried to the nervous membranes of the body and joints, causing the inflammatory condition incident to rheumatism.

Ginseng combined with the juices of a good ripe pineapple is par excellent as a treatment for indigestion. It stimulates the healthy secretion of pepsin, thereby insuring good digestion without incurring the habit of taking pepsin or after-dinner pills to relieve the fullness and distress so common to the American people. The above compound prepared with good wine in the proper way will relieve many aches and pains of a rebellious stomach; and if I should advise or prescribe a treatment for the old "sang digger" who is troubled with dyspepsia or foul stomach, I would tell him to take some of your own medicine and don't be selling all to the Chinamen.

[Illustration: A Healthy Looking "Garden"--"Yard."]

I want to repeat here what I have often said to "sengers" of my acquaintance, especially those "get-rich-quick" fellows who have been dumping their half-grown and poorly cured Ginseng on the market, thereby killing the good-will of the celestial for a market and destroying the sale of those who cultivate clean and matured roots; they had much better give their roots time to mature in their gardens and if the market price is not what it ought to be to compensate for the labor, they had better hold over another season before selling. I have all the product of last season in Ginseng and Golden Seal in my possession, for the reason that the price did not suit me. Drug manufacturers ask

CHAPTER XII.

$7.00 per pound for Fluid Extract Golden Seal wholesale. When they can make from one-half pound dried root one pound Fluid Extract Golden Seal costing them 75 cents, that's a pretty good profit for maceration and labeling.

Ginseng has been used to some extent as a domestic medicine in the United States for many years. As far as I can learn, the home use is along the line of tonic and stimulant to the digestive and the nervous system. Many people have great faith in the power of the Ginseng root to increase the general strength and appetite as well as to relieve eructations from the stomach. As long ago as Bigelow's time, some wonderful effects are recorded of the use of half a root in the increase of the general strength and the removal of fatigue. Only the other day a young farmer told me that Ginseng tea was a good thing to break up an acute cold and I think you will find it used for rheumatism and skin diseases. It undoubtedly has some effect on the circulation, perhaps thru its action on the nervous system and to this action is probably due its ascribed anti-spasmodic properties.

The use of Ginseng has largely increased within the last few years and several favorable reports have been published in the medical journals. One physician, whose name and medium of publication I cannot now recall, speaks highly of its anti-spasmodic action in relieving certain forms of hiccough. If this is true, it places it at once among the important and powerful anti-spasmodics and suggests its use in other spasmodic and reflex nervous diseases as whooping cough, asthma, etc.

I have practiced medicine for eight years. I sold my practice one year ago and since have devoted my entire attention to the cultivation of Ginseng and experimenting with Ginseng in diseases and am satisfied that it is all that the Chinese claim for it; and, if the people of the United States were educated as to its use, our supply would be consumed in our own country and it would be a hard blow to the medical profession.

It would make too long an article for me to enumerate the cases that I have cured; but, I think it will suffice to say that I have cured every case where I have used it with one exception and that was a case of consumption in its last stages; but the lady and her husband both told

CHAPTER XII.

me that it was the only medicine that she took during her illness that did her any good. The good it did her was by loosening her cough; she could give one cough and expectorate from the lungs without any exertion. I believe it is the best medicine for consumption in its first stages and will probably cure.

I wish the readers of Special Crops to try it in their own families--no difference what the disease is. Make a tea of it. A good way is to grate it in a nutmeg grater. Grate what would make about 15 grains, or about one-fourth to one-half teaspoonful and add half a pint or less of boiling water. The dose to be taken at meal times and between meals. In a cold on the lungs it will cure in two or three days, if care is taken and the patient is not exposed.

My theory is that disease comes from indigestion directly or indirectly. Ginseng is the medicine that will regulate the digestion and cure the disease no difference by what name it is called; if the disease can be cured. Ginseng will cure it where no other drug will.

I will cite one case; a neighbor lady had been treated by two different physicians for a year for a chronic cough. I gave her some Ginseng and told her to make a tea of it and take it at meal times and between meals; in two weeks I saw her and she told me that she was cured and that she never took any medicine that did her so much good, saying that it acted as a mild cathartic and made her feel good. She keeps Ginseng in her house now all the time and takes a dose or two when she does not feel well.

I am satisfied that wonderful cures can be made with Ginseng and am making them myself, curing patients that doctors have given up; and if handled properly our supply will not equal the demand at home in course of five or six years, thus increasing the price.

* * *

At the last annual meeting of the Michigan Ginseng Association, Dr. H. S. McMaster of Cass Co. presented a paper on the uses of this plant, which appeared in the Michigan Farmer. He spoke in part as follows:

CHAPTER XII.

"Ginseng is a mild, non-poisonous plant, well adapted to domestic as well as professional uses. In this respect it may be classed with such herbs as boneset, oxbalm, rhubarb and dandelion. The medicinal qualities are known to be a mild tonic, stimulant, nervine and stomachic. It is especially a remedy for ills incident to old age.

"Two well-known preparations made--or said to be--from Ginseng root are on the market. One of these, called "Seng," has been for many years on druggists' shelves. It is sometimes used for stomach troubles and with good results. I think it is now listed by the leading drug houses.

"Another called 'Ginseng Tone' is a more recent preparation, and is highly spoken of as a remedy. But for home or domestic use we would suggest the following methods of preparing this drug:

"1st. The simplest preparation and one formerly used to some extent by the pioneers of our forest lands, is to dig, wash and eat the green root, or to pluck and chew the green leaves. Ginseng, like boneset, aconite and lobelia, has medicinal qualities in the leaf.

"To get the best effect, like any other medicine it should be taken regularly from three to six times a day and in medicinal quantities. In using the green root we would suggest as a dose a piece not larger than one to two inches of a lead pencil, and of green leaves one to three leaflets. These, however, would be pleasanter and better taken in infusion with a little milk and sweetened and used as a warm drink as other teas are.

"2nd. The next simplest form of use is the dried root carried in the pocket, and a portion as large as a kernel of corn, well chewed, may be taken every two or three hours. Good results come from this mode of using, and it is well known that the Chinese use much of the root in this way.

"3d. Make a tincture of the dried root, or leaves. The dried root should be grated fine, then the root, fiber or leaves, separately or together, may be put into a fruit jar and barely covered with equal parts of alcohol and water. If the Ginseng swells, add a little more alcohol and

CHAPTER XII.   98

water to keep it covered. Screw top on to keep from evaporating. Macerate in this way 10 to 14 days, strain off and press all fluid out, and you have a tincture of Ginseng. The dose would be 10 to 15 drops for adults.

"Put an ounce of this tincture in a six-ounce vial, fill the vial with a simple elixir obtained at any drug store, and you have an elixir of Ginseng, a pleasant medicine to take. The dose is one teaspoonful three or four times a day.

"The tincture may be combined with the extracted juice of a ripe pineapple for digestion, or combined with other remedies for rheumatism or other maladies.

"4th. Lastly I will mention Ginseng tea, made from the dry leaves or blossom umbels. After the berries are gathered, select the brightest, cleanest leaves from mature plants. Dry them slowly about the kitchen stove in thick bunches, turning and mixing them until quite dry, then put away in paper sacks.

"Tea from these leaves is steeped as you would ordinary teas, and may be used with cream and sugar. It is excellent for nervous indigestion.

"These home preparations are efficacious in neuralgia, rheumatism, gout, irritation of bronchi or lungs from cold, gastro-enteric indigestion, weak heart, cerebro-spinal and other nervous affections, and is especially adapted to the treatment of young children as well as the aged. Ginseng is a hypnotic, producing sleep, an anodyne, stimulant, nerve tonic and slightly laxative."

CHAPTER XIII.

GINSENG IN CHINA.

With the exception of tea, says the Paint, Oil and Drug Review, Ginseng is the most celebrated plant in all the Orient. It may well be called the "cure-all" as the Chinese have a wonderful faith in its curative and strengthening properties, and it has been appropriately called the "cinchona of China." It is considered to be a sovereign cure for fevers and weaknesses of all kinds, and is, indeed, the chief and most costly medicine of the Chinese Empire.

Ginseng is found wild in the mountain forests of eastern Asia from Nepa to Manchuria. It once grew in Fukien, Kaighan and Shansi, but was supplanted by the Manchuria wild root. The root is carefully hunted for by the Manchus, who boast that the weeds of their country are the choice drugs of the Chinese, a boast which has much foundation in fact. Of the thirty-seven ports in China where the imperial maritime customs are established to import Ginseng, imports during 1905 were as follows: Shanghai, 103,802 pounds; Wuhu, 2,374; Kiuhiang, 2,800; Hankow, American clarified, 34,800; Wenchau 9,100; Chungking, American clarified, 6,200; Chefoo, 80,408; Canton, 75,800, and Foochow, 15,007.

The total importation at these ports for the last four years were: 1902, 407,021 pounds; 1903, 404,000 pounds; 1904, 313,598 pounds, and 1905, 331,381 pounds. These figures, however, by no means cover all the Ginseng entering China, as much of it comes thru the native custom houses, which keep no tabulated data of exports and imports, and great quantities of it are smuggled into the country, especially over the Korean boundary line. Niuchwang is the one Chinese port which exports native Ginseng. Its exports for the last four years were, respectively, 228,000, 215,000, 57,000 and 160,900 pounds.

To give an accurate price for Ginseng would be impossible, so greatly does it differ from the variety of the root offered to consumers. Some wild roots have been known to realize their weight in gold; while the cultivated variety can be purchased from 5 cents a pound up. Generally speaking, the present average prices are, for the best

Ginseng, $12.00 a pound; for fair quality, $6.50, and for the ordinary, 50 cents to $1.00. Japan sends to China the cheapest Ginseng, a great deal of which is used to adulterate the highest quality from Korea.

In values and quality of the root the four principal producing countries rank as follows: Manchuria, Korea, America and Japan. Prices often vary in accordance with the method used in clarifying the root. Some Chinese provinces prefer it white, others reddish and still others require it of a yellowish tinge. The Korean root is reddish in color, due, some say, to the ferruginous soil on which it grows, and, according to others, to a peculiar process of clarifying. Most of the Korean product goes to southern China by way of Hongkong.

Wild Ginseng, from whatever country, always commands a better price than the cultivated article, chiefly because of Chinese superstition, which prefers root resembling man or some grotesque creature to that of the regular normal roots which cultivation naturally tends to produce. Chinese druggists, when questioned as to the real difference between the Manchuria wild and the American cultivated Ginseng root, admit that the difference in quality is mostly imaginary, altho there is a real difference in the appearance of the roots.

But the Manchuria Ginseng comes from the Emperor's mother country and from the same soil whence sprang the "god of heaven" and therefore the Chinese regard it as infinitely more efficacious as a curative agent than any other Ginseng could possibly be. Many assert that the future demand for Ginseng will be a decreasing one, from the fact that its imaginary properties of curing every disease on earth will be dissipated in proportion to the advance of medical science. There can be no doubt, however, that Ginseng does possess certain curative properties and it can be safely asserted that it will require many generations, perhaps centuries, to shake the Chinaman's faith in his mysterious time-honored cure-all.

[Illustration: Root Resembling Human Body.]

American Ginseng, of which large quantities are annually exported to China, is classed, as a rule, with hsiyang, that is, west ocean, foreign

CHAPTER XIII. 101

or western country Ginseng. The imports of this article at Niuchwang for 1905 amounted in value to $4,612 gold. The exports of Manchurian Ginseng thru Niuchwang to Chinese ports for 1905 aggregated in value $180,199 gold and for 1904, $205,431 gold. Wild Manchuria Ginseng is rare, even in Manchuria, and its estimated valuation ranges at present from $450 to $600 gold a pound.

The total imports of Ginseng into China for 1904 aggregated 277 tons, valued at $932,173.44 and for 1905 to 1,905 tons, valued at $1,460,206.59. The increased valuation of the imports of last year emphasizes the increased price of Ginseng in the Chinese market.

Hsiyang, or American Ginseng, is marketed in China largely thru Hongkong and Shanghai foreign commission houses. Importations of the American product are increasing in bulk with each succeeding year, and the business gives every indication of becoming a very large one in a short time.

* * *

In most of the booklets and articles we have seen on Ginseng, the writers quote exorbitant figures as to what the root sells for in China. A good many of them quote from reports received from U. S. Consuls, who, when they give prices, reckon on Mexican dollars which are only about half the value of ours and some of them go so far as to quote retail prices for very small quantities of extra quality root.

Some of the growers and dealers in this country, therefore, imagine that they are not paid what they should be for their stock and that there is an enormous profit for the men who ship to China. Such is an entirely wrong idea and can be best proven by the fact that during the past couple of years three of the leading export houses have gone out of business, owing to there being no money in it. We do not know of any business conducted on as small a percentage profit as Ginseng. Frequently prices paid in this country are in excess of the market in China.

This not only means a direct loss to the exporter on his goods but also the cost of making clean (removing fibres, siftings and stems) shrinkage, insurance and freight. Business is also conducted on different lines from years ago. Then the buyers in China bought readily, prices were lower and more people could afford to use it.

Today, prices are tripled and while the supply is smaller, the demand is very much less and Chinese buyers make the exporters carry it until they really need it, in a good many cases buying root and not taking it for three or four months, and consequently keeping the exporters without their money. The expense of carrying Ginseng is also heavy owing to the high rate of interest, which is 8% and over.

The folly of depending upon U. S. Consul reports is shown in the great difference in the figures which they send. Many of these men have but very little knowledge of business, most of them knowing more about politics. It is not likely that this class of men will spend very much time in investigating a subject of this character.

The market here for wild root since June 1st has been the dullest we have ever known and the same condition prevails in China. We are glad to state that cultivated root is selling at much better prices than last year. It is hard to account for the disfavor with which it was regarded a year ago in China, and the prejudice against it has been overcome more rapidly than we expected. At this time last year it was almost unsalable and we were buying as low as $3.00 to $4.00 per pound. Many houses declined to buy at all.

Now that the prejudice against it has sort of worn off, we look for a good market and consider the outlook very favorable and would advise people not to give up their gardens in too great a hurry. We make a specialty of cultivated root and will be pleased to give information as to handling, drying, etc., to any reader who desires it. We have been buying Ginseng for over thirty years.

Belt, Butler Co. New York.

* * *

Consul-General Amos P. Wilder of Hongkong, in response to numerous American inquiries as to the trade in Ginseng, with especial

CHAPTER XIII.

reference to the cultivated root, prices and importations, reports as follows:

The Ginseng business is largely in the hands of the Chinese, the firms at Hongkong and Canton having American connections. (The five leading Hongkong Chinese firms in the Ginseng importing business are named by Mr. Wilder, as also the leading "European" importing concern, and all the addresses are obtainable from the Bureau of Manufacturers).

I am authorized to say that American growers may correspond with the European concern direct relative to large direct shipments. They receive goods only on consignment and have some forty years' standing in this industry. This firm, as do the Chinese, buys in bulk and distributes thru jobbers to the medicine shops, which abound in all Chinese communities. The Cantonese have prestige in cleaning and preparing the root for market.

Last year the best quality of Ginseng brought from $2,000 to $2,300 Mexican per picul (equal to 133 1/2 pounds), but selected roots have brought $2,400 to $2,550. It is estimated here that growers should net about $7.25 gold per pound. The buying price of Ginseng is uncertain. There being no standard, no price can be fixed. The American-Chinese shippers have the practice of withholding the Ginseng to accord with the demand in China. Owing to failures among Chinese merchants since the war and the confusion in San Francisco, trade in this industry has been slack and prices have fallen off. If the root is perfect and unbroken it is preferred. Much stress should be laid on shipping clean, perfect and attractive roots. Size, weight and appearance are factors in securing best prices, the larger and heavier the root the better.

When the shipment arrives the importer invites jobbers to inspect the same. The roots are imported in air-tight casks in weight of about 100 pounds. It is certain that there are many different qualities of Ginseng and the price is difficult to fix (except on inspection in China).

As to wild and cultivated roots, two or three years ago when cultivated Ginseng was new, buyers made no distinction and the price ruled the

CHAPTER XIII.   104

same; but having learned of the new industry, experts here assure me the roots can readily be distinguished. They say that the wild root is darker in color and rougher. The wild is preferred. Experts now allege a prejudice against the cultivated root, affirming that the wild root has a sweeter taste. The cultivated roots being larger and heavier, they first earned large prices, but are now at a disadvantage, altho marketable.

[Illustration: Wild Ginseng Roots.]

The cultivated is as yet but a small percentage of the entire importations, but is increasing. Seventy-five per cent of all importations are in the hands of the Chinese. Small growers in America will do best to sell to the collecting buyers in New York, Cincinnati and other cities. Hongkong annual importations are now about 100,000 pounds.

Too many misleading and conflicting articles have been published on the subject of Ginseng culture in Korea, a true statement of the facts may be of interest. We all know the Korean Ginseng always commands a high price in China and I believe there must be a very good reason for it. Either the Korean method of cultivation, curing or marketing was superior to the American method or centuries of experience in its cultivation had taught him a lesson and a secret we had yet to learn. After considerable correspondence with parties in Korea which gave me very little information and to set my mind at rest on these questions, I went to Korea in 1903 for the sole purpose of obtaining all the information possible on Ginseng culture according to Korean methods and also if possible to secure enough nursery stock to plant a Ginseng garden in America with the best Korean stock.

Strange to say, even after I reached the city of Seoul, the capital of Korea, I could not obtain any more reliable information on Ginseng than I already knew before I left America. They told me where the great Ginseng district was located, that 40,000 cattys were packed each year for export, etc., but as to the soil, planting, cultivation, irrigation, shading, curing, packing, etc., they knew nothing that was reliable.

All the American people use sugar in one form or another, but how many could tell a person seeking for reliable information concerning

CHAPTER XIII.

the planting of the cane or sugar beet, of the character of the soil necessary, of its cultivation and irrigation, the process of refining, packing and marketing, etc. Comparatively few, indeed, and so it is with the Koreans on the cultivation of Ginseng. They all use it, but, like the Chinese, not one in several thousand ever saw a Ginseng plant growing. After considerable delay I secured a competent interpreter, a cook, and food supplies, and started from Seoul for the great Ginseng district, traveling part of the way by rail, then by sampan, and finally reached my destination on Korean ponies. Arriving at the Ginseng center, I lived among the Ginseng growers from the time the seed crop ripened until nearly all the five-year-old roots, or older ones, were dug up and delivered to the government at their drying grounds, which is about four acres in extent. This compound is enclosed on three sides by buildings from 100 to 150 feet in length and a uniform width of twelve feet and the rest of the compound with a high stone wall with a gate, which is closely guarded by soldiers armed with guns. Near the center of this compound is a well where the roots are washed as soon as they are received. There is no entrance from the outside to any of these buildings. Every one must pass the guards at the gate, for the buildings, together with the wall, make a complete enclosure.

The Ginseng gardens are scattered over considerable territory, most of which is surrounded by a high stone wall about twenty or twenty-five miles in circumference, similar to the great wall of China, and which many years ago was the site of one of the ancient capitals of Korea.

Part of the growers make a specialty of raising one-year-old plants, to supply those who have sufficient means to wait four years more for the roots to mature. Generally, speaking, the grower that produces the commercial root raises but little if any one-year roots.

All Ginseng gardens are registered as required by law, stating how many kan (a kan of Ginseng is the width of the bed, about 30 inches and 5 1/2 feet long) are under cultivation, so the High Government Official, specially appointed for the Ginseng district, always knows how many roots should be available at harvest time and every grower must sell his entire crop that is five years old or over to the government and his responsibility does not cease until he has delivered his crop at the government drying grounds.

CHAPTER XIII.

His roots are then carefully selected and all that do not come up to a required size are rejected and delivered back to the grower and these he can either dry for his own use or he can transplant them and perhaps next year they will come up to the required standard. The Koreans pay great attention to the selection of their Ginseng seed. No plant is allowed to bear seed that is less than four years old and very little seed is used from four-year plants. Nearly all the seed comes from five-year-old plants and a little from six-year-old. Only the best and strongest appearing plants are allowed to bear seed, and even these very sparingly, as part of the seed head is picked off while in the blossom and from which they make a highly prized tea. The seed stem of all other plants are pinched off, forcing all the strength, as well as medicinal properties, into the root.

Many of the best growers never allow their plants to bear seed, and only the required amount of seed is raised each year to supply the demand. After the seed is gathered, it is graded by passing it thru a screen of a certain size. This grader is made like an old-fashioned flour sieve, only the bottom is made of a heavy oil paper with round holes cut in it, and all seed that will pass thru these holes are destroyed, so only the largest and best seed are kept for planting. The soil which they use for their Ginseng garden is a very poor disintegrate granite, to which has been added leaf mould mostly from the chestnut oak, in the proportion of three-eighths leaf mould to five-eighths granite. The leaves are gathered in the spring and summer, dried in the sun, pulverized and sprinkled with water to help decomposition. This is the only fertilizer used. The beds are raised about eight inches above the level of the ground and are carefully edged with slabs of slate. What is called a holing board is used to mark the places for the seed. It is made of a board as long as the beds are wide (about thirty inches) and has three rows of pegs 1/2-inch long and 1 1/2 inches apart each way.

A seed is planted in each hole and covered by pressing the soil down with the hands. About 1/4-inch of prepared soil is added to the bed and smoothed over. No other mulch is used. The roots are transplanted each year, setting them a little farther apart each time, until at the third transplanting, or at four years old, they are 6x6 inches apart, and at each transplanting the amount of leaf mould used in the prepared soil is reduced. (Note the difference between this and the American

CHAPTER XIII. 107

method of heavy fertilizing). Only germinated seed is planted and the time for planting is regulated by the Korean Calendar and not by the weather and if at that time it is at all cold, the beds are immediately covered with one or two thickness of rice straw thatch and as soon as the weather is suitable this thatch is removed and the shade erected. Each bed is shaded separately by setting a row of small posts in the ground 4 feet high and 5 1/2 feet or 1 kan apart, on the north side of each bed and on the south side a similar row, only about 1 foot high. Bamboo poles are securely lashed to these posts and they in turn support the cross pieces on which rests the roof covering, made of reeds woven together with a very small straw rope. At the time of the summer solstice, the rainy season comes on, so a thick covering of thatch is spread over the reed covering, which sheds the rain into the walks, while the back and front are enclosed with rush blinds, that on the north side being raised or lowered according to the temperature. If it is a very hot day the blinds are lowered from about 10 A. M. to 4 P. M., leaving the beds in almost darkness.

The beds are all protected from the rain and are irrigated by sprinkling them when needed. At the close of the growing season, after the roots have gone dormant, all that are not dug up are covered with a layer of soil 7 or 8 inches thick. All the shade is pulled down except the posts and spread over the soil and the garden is left thus for the winter, and the grower selects another site to which he can move his plants in the spring, and each year new soil is prepared. From the time the roots are two years old there is another added care. They are now worth stealing, consequently the garden has to be watched day and night. A watch tower about 16 feet high is erected and the hands take turn about, occupying it as a sentry. Another man constantly patrols the garden during the night.

The Koreans are the largest consumers of Ginseng in the world, in proportion to their population, and they have carefully cultivated it for centuries with the one particular object in view, "its medicinal properties." For quality always, rather than quantity. They sacrifice everything else for a powerful medicinal root, and they surely grow it. I have seen some remarkable results from its use during my stay in Korea. Say what we may about it, but it plays a very important part in the life of both the Korean and the Chinese people. Do you wonder

CHAPTER XIII.

now that the Korean Ginseng always commands a high price? If the American growers had followed closer along the lines of the Korean growers and aimed for a high grade of medicinal root, the market for American Ginseng would not be where it is today. That is, the cultivated Ginseng. The American growers have it in their own hands to either make a success or failure of Ginseng culture, but one thing is certain, heavy seed bearing, excessive fertilizing and rapid drying will never produce a high standard of Ginseng. The principal market of the world is ours if we only reach out for it with that high standard and maintain it and especially so if we will unite together and market our product thru one central agency controlled by the producers. Mr. Chinaman may sometimes be mistaken as to whether Ginseng is wild or cultivated. He may also be mistaken as to whether it comes from Korea or China (I have seen him make this mistake), but let him once sample a liberal dose of it, and he won't make any mistake as to whether it is good, medium or bad.

* * *

The Ginseng Trade.

The following article by Mr. Burnett appeared in the Minneapolis Journal last February and shows what dealers think of the Ginseng industry:

I wish you would give room for what I have to say in regard to an article in your Journal last fall by our ex-Consul, John Goodnow. Some things he says are correct: That the demand is based entirely on superstition; that the root has life-giving qualities; and that those having the nearest resemblance to human beings are most valuable. That is quite true. I have seen the Chinese exporters' eyes dance when they saw such roots in a lot.

Now for the errors in what he said. He says the trade is in the hands of a syndicate and they only handle Korean Ginseng. Possibly this syndicate tells the Chinese retail merchants that to keep them from boycotting our American Ginseng. If so, why is it that the wild root this fall has been at ready sale at $6.75 to $7.10 per pound? We, who buy it, do not hold it and if we did not find a ready sale for it we would soon

CHAPTER XIII.

cease to buy it.

There has been marketed in Minneapolis probably $50,000 worth this year and in the United States a million dollars' worth. So you see his error: for, either directly or indirectly, it gets to China at good prices.

Chinese Superstitions.

Now in regard to the cultivated root, to show your readers how the value is based on superstition, we will cite one instance in our experience. We sent our clerk to a laundry where there were a half dozen "Celestials" to sell some nice cultivated root. Some roots were manlike in shape. They tasted it, were delighted with it and bought it readily and told him to bring them all he could get, as what they did not need for their own use they would ship to their exporter in San Francisco.

Our man told them he would be around in one week. We sent him again in just a week. He said on his return they "looked daggers" at him and said, "We no wantee your cultivated root." This convinced us they had shipped it to the agents of the syndicate at 'Frisco and received their returns. Now, does this not show that the demand is all based on superstition? It was very good until they were informed that it was cultivated.

Now your readers may say, how can they distinguish between the cultivated and the wild? I will tell you; the cultivated is usually much firmer and twice as heavy as the wild and generally much cleaner. Then most of the cultivated has been raised from small, wild roots dug from the forests and in transplanting they have not taken pains to place the tap root straight in the earth. This causes it to be clumpy--that is, not straight like most wild roots. This, with its solidity and cleanliness makes it easy to tell from the wild roots.

[Illustration: Pennsylvania Grower's Garden.]

The Cultivated Plant.

CHAPTER XIII.

Now we have had a number of lots of cultivated that we got full prices for. They were roots grown from seeds, symmetrical in shape, not too large, not too clean and dug before they became very solid. My idea is, if not allowed to grow more than as large as one's fingers, when dry and dug immediately after the seeds are ripe, or even before, if seeds are not needed, and not washed too clean, we can find sale for such. At present the ordinary cultivated does not bring quite half the price of the wild. There are some who buy that for American use, several firms putting up Ginseng cures. Some people, like the Chinese, believe it has merits, but as the demand is limited the price is low. That the Chinese think that the root grown by nature has life-giving qualities and that cultivated has no virtues, is certain. The only way to do is to grow in natural woods soil (manure of any kind must be avoided, as it causes a rank growth) dig and wash it so they can't tell the difference. One thing is certain, it's a hardy plant, altho slow to get started, and good money can be made at $2.00 to $3.00 a pound. Instead of being hard to grow, as many persons think, it is very hard to kill.

* * *

A belief among the Chinese people is that Ginseng roots, especially if of peculiar shape, will cure practically all diseases of mind and body. The Chinese are not given to sentiment; their emotional nature is not highly developed; they are said to be a people who neither "kiss nor cuss," and their physical sensibilities are so dull that a Chinaman can lie down on his back across his wheelbarrow with feet and head hanging to the ground, his mouth wide open and full of flies and sleep blissfully for hours under the hottest July sun. There is nothing about them, therefore, to suggest that they possess the lively imagination to make them have faith in a remedy with purely imaginary virtues. Nevertheless, among these people, a plant not found by any medical scientist to possess any curative powers is used almost universally, to cure every kind of ailment and has been so used for generations.

Intelligent Chinese resent the imputation of superstition to their people. But the fact remains that the Ginseng roots are valued according to the peculiarity of their shapes. The word Ginseng is composed of two Chinese words which mean man and plant, and the more nearly shaped like a man the roots are, the more they are valued. A root which is bifurcated and otherwise shaped like a man, may be sold as high as $10.00 an ounce; a recent secretary of the Chinese Legation explains this on the ground of being valued as a curio; but the curio is finally made into a decoction and swallowed, and the swallower evidently hopes that the fantastic shape of the root will make the medicine more potent.

CHAPTER XIV.

GINSENG--GOVERNMENT DESCRIPTION, ETC.

The following is from a bulletin issued by the U. S. Department of Agriculture--Bureau of Plant Industry--and edited by Alice Henkel:

Panax Quinquefolium L.

Other Common Names--American Ginseng, sang, red-berry, five-fingers.

Habitat and Range--Ginseng is a native of this country, its favorite haunts being the rich, moist soil in hardwood forests from Maine to Minnesota southward to the mountains of northern Georgia and Arkansas. For some years Ginseng has been cultivated in small areas from central New York to Missouri.

Description of Plant--Ginseng is an erect perennial plant growing from 8 to 15 inches in height and bearing three leaves at the summit, each leaf consisting of five thin, stalked ovate leaflets, long pointed at the apex, rounded or narrow at the base, the margins toothed; the three upper leaflets are largest and the two lower ones smaller. From 6 to 20 greenish yellow flowers are produced in a cluster during July and August, followed later in the season by bright crimson berries. It belongs to the Ginseng family (Araliaceae.)

Description of Root--Ginseng has a thick, spindle-shaped root, 2 to 3 inches long or more, and about one-half to 1 inch in thickness, often branched, the outside prominently marked with circles or wrinkles. The spindle-shaped root is simple at first, but after the second year it usually becomes forked or branched, and it is the branched root, especially if it resembles the human form, that finds particular favor in the eyes of the Chinese, who are the principal consumers of this root.

[Illustration: Ginseng (Panax Quinquefolium).]

Ginseng root has a thick, pale yellow white or brownish yellow bark, prominently marked with transverse wrinkles, the whole root fleshy and

CHAPTER XIV.

somewhat flexible. If properly dried, it is solid and firm. Ginseng has a slight aromatic odor, and the taste is sweetish and mucilaginous.

Collection and Uses--The proper time for digging Ginseng root is in autumn, and it should be carefully washed, sorted and dried. If collected at any other season of the year, it will shrink more and not have the fine, plump appearance of the fall dug root.

The National Dispensatory contains an interesting item concerning the collection of the root by the Indians. They gather the root only after the fruit has ripened, and it is said that they bend down the stem of ripened fruit before digging the root, covering the fruit with earth, and thus providing for future propagation. The Indians claim that a large percentage of the seeds treated in this way will germinate.

Altho once official in the United States Pharmacopoeia, from 1840 to 1880, it is but little used medicinally in this country except by the Chinese residents, most of the Ginseng produced in this country being exported to China. The Chinese regard Ginseng root as a panacea. It is on account of its commercial prominence that it is included in this paper.

Cultivation--There is probably no plant that has become better known, at least by name, during the past ten years or more than Ginseng. It has been heralded from north to south and east to west as a money-making crop. The prospective Ginseng grower must not fail to bear in mind, however, that financial returns are by no means immediate. Special conditions and unusual care are required in Ginseng cultivation, diseases must be contended with, and a long period of waiting is in store for him before he can realize on his crop.

Either roots or seeds may be planted, and the best success with Ginseng is obtained by following as closely as possible the conditions of its native habitat. Ginseng needs a deep, rich soil, and being a plant accustomed to the shade of forest trees, will require shade, which can be supplied by the erection of lath sheds over the beds. A heavy mulch of leaves or similar well rotted vegetable material should be applied to the beds in autumn.

CHAPTER XIV.

If roots are planted, they are set in rows about 8 inches apart and 8 inches apart in the row. In this way a marketable product will be obtained sooner than if grown from seed. The seed is sown in spring or autumn in drills 6 inches apart and about 2 inches apart in the row. The plants remain in the seed bed for two years and are then transplanted, being set about 8 by 8 inches apart. It requires from five to seven years to obtain a marketable crop from the seed. Seed intended for sowing should not be allowed to dry out, as this is supposed to destroy its vitality.

Price--The price of wild Ginseng roots ranges from $5.00 a pound upward. The cultivated root generally brings a lower price than the wild root, and southern Ginseng roots are worth less than those from northern localities.

Exports--The exports of Ginseng for the year ended June 30, 1906, amounted to 160,949 pounds, valued at $1,175,844.

CHAPTER XV.

MICHIGAN MINT FARM.

Very few people know that the largest Mint farm in the world is owned and operated by an unassuming Michigan man named A. M. Todd, says Special Crops. His career is interesting. Born on a farm near St. Joseph, Mich., he early developed an idea that money was to be made in the growing of Peppermint. At that time the Mint oil industry was small and in a state of crudeness in America, for Europe was supposed to be the stronghold of the industry. To Europe went Mr. Todd to see about it. He returned filled with plans and enthusiasm.

Some Details of the Business.

The details are long, but the main facts can be briefly told. Eventually, while still a very young man, Mr. Todd purchased 1,400 acres of wild, swampy land in Allegan County, Mich. The purchase price was $25,000. He proceeded to hire a force of men to clear and ditch the new Mint farm. That was 20 or more years ago.

Now, let us take a look at that farm as it is today. First we come to the main farm, called Campania, and comprising just 1,640 acres. Here are huge barns, comfortable houses for employer and employees, warehouses, ice houses, windmills, library, club rooms and bathrooms for use of employes; 17 miles of wide, deep, open drainage ditches; stills for distilling Peppermint oil; roadways, telephones and all the system and comfort of a little village founded and maintained by one thoughtful man.

Not far away is a second farm, recently purchased where somewhat similar improvements are now going on. This farm is named Mentha, and consists of 2,000 acres.

Then, farther north, a third farm completes the Todd domain. This place contains 7,000 acres and is known as Sylvania Range. The three farms, with a total acreage of 10,640 acres, are under one management and they form together the largest Mint farm in all the world. Starting with $100.00 capital, Mr. Todd's plant today is worth several hundred thousand dollars.

Distiller as Well as Grower.

But Mr. Todd is more than a Mint grower. With his distilleries he turns the crop into crude Peppermint oil; with his refineries he turns the crude oil into the refined products that find a ready market in the form of menthol, or as a flavoring essence for drinks, confectionery and chewing gum, or for use in medicine. Furthermore, he has been shrewd enough to figure out a method of utilizing, profitably, the by-products of the business, Mint hay. In other words, after the oil is extracted from a mass of Mint plants in a distillery vat, the resulting cake of leaves and stems is dried and fed to cattle. And, oddly enough, the animals greatly relish it and thrive upon it.

Raises Shorthorns on Mint Hay.

During the summer Mr. Todd has 500 Shorthorns grazing on his 7000-acre range, where they require no human attention during the season when his men are busy planting, cultivating and harvesting the first crop. Later, these same Shorthorns are driven from pasture to the big Campania barns, where the men care for them and feed them Mint hay from Mr. Todd's distilleries at a season when such workmen have little else to do. In this way the by-product is utilized and the regular force of men is kept employed all the year around.

The growing of Mint is simple, yet there are some peculiar features about it. For instance, the land is so shaky at some seasons of the year that horses can not work on it unless they wear special, broad wooden shoes. This Mint soil, indeed, is something like the muck found in typical celery fields, being black, damp and loose. But it is less firm and more damp than the celery land at Kalamazoo.

Setting New Mint Fields.

The Mint root is perennial. Once in two or three years, however, the fields are renewed to improve the crop. When setting a new field the land is plowed and harrowed in the usual way. It is then marked out in shallow furrows into which the sets are evenly dropped by skilled

CHAPTER XV.

planters who cover each dropped root by shoveling dirt over it with the foot. The rows are about 2 1/2 feet apart and the planting is done in early spring. The sets are obtained by digging up and separating the runners and roots from old plants.

The planted rows soon send up shoots above ground and the new plants rapidly run or spread, necessitating hoeing and cultivating only until late July, at which time the field should be densely covered with a rank growth of waving green plants that forbid further cultural work.

Harvesting the Mint.

In August or September the field is mowed, raked and bunched; in fact, handled quite similarly to a clover hay field. After allowing the plants to dry a short time, the crop is loaded onto hay wagons and carted to the stills, where the essential oil is extracted by means of a system of steam distillation.

The second year's crop is obtained by the simple method of plowing under the plants in the fall. The roots send up new shoots next season, while weeds are temporarily discouraged. No cultivation is attempted the second year, altho the hand pulling of weeds may sometimes prove desirable.

We think the growing of Mint should not be attempted except on a large scale. We have had many queries touching the plant and manner of cultivation that we have taken this means to answer them. In boyhood days we were well acquainted with this industry in all its branches and can not advise the average Ginseng grower to undertake its culture for the reason that there is not money enough in it to be profitable on small areas of land.

CHAPTER XVI.

MISCELLANEOUS INFORMATION.

Remember, unless thoroughly dried roots, herbs, leaves, barks, flowers and seeds are apt to heat or mold which greatly lessens their value. If badly molded they are of little value.

The best time to collect barks is in the spring (when the sap is up) as it will peel easier at that time. Some barks must be rossed, that is, remove the outer or rough woody part. In this class are such barks as white pine, wild cherry, etc.

Leaves and herbs should only be gathered when the plant is mature-grown. In curing they should be kept from the sun as too rapid curing tends to draw the natural color and this should be preserved as much as possible.

Flowers should be gathered in the "height of bloom," for best results. They require considerable attention to preserve as they are apt to turn dark or mold.

The time to gather seeds is when they are ripe. This can easily be determined by the leaves on the plant, vine or shrub which produced the seeds. Generally speaking, seeds are not ripe until early fall, altho some are.

There has been a heavy demand for years for wild cherry bark, sassafras bark, black haw bark, prickly ash bark, slippery elm bark, cotton root bark as well as scullcap plants, (herbs) lobelia herb, golden thread herb and red clover tops.

There has been a cash market for years for the following roots: Blood, senega, golden seal, poke, pink, wild ginger, star, lady slipper, black, mandrake, blue flag and queen's delight.

If you have a few pounds of Ginseng or Golden Seal, pack carefully in a light box and ship by express. If less than four pounds, you can send by mail--postage is only one cent an ounce. A four-pound package by

CHAPTER XVI.

mail can be sent anywhere in America for 64 cents. Expressage, unless short distances, is apt to be more.

[Illustration: Lady Slipper.]

In shipping roots, herbs, leaves, seeds, etc., where the value is only a few cents per pound it is best to collect 50 pounds or more before making a shipment. In fact, 100 pounds by freight costs no more than 10, 20, 50 or any amount less than 100 as 100 pounds is the smallest charge.

Some of the biggest liars in America seem to be connected with the "seng" growing business. They probably have seed or plants to sell. Be careful in buying--there are many rascals in the business.

There is always a cash market for Ginseng and Golden Seal. In the large cities like New York, Chicago, St. Louis, Minneapolis, Montreal, Cincinnati, etc., are dealers who make a special business of buying these roots. In hundreds of smaller cities and towns druggists, merchants, raw fur dealers, etc., buy them also. The roots, barks, leaves, etc., of less value are also bought pretty generally by the above dealers, but if you are unable to find a market for them it will pay you to send 10 cents for copy of Hunter-Trader-Trapper, Columbus, Ohio, which contains a large number of root buyers' advertisements as well as several who want bark, leaves, seeds, flowers, herbs, etc.

Since 1858 Ginseng has increased in value one thousand four hundred per cent., but Golden Seal has increased in value in the same time two thousand four hundred per cent.

Ginseng and Golden Seal should be packed tightly--light but strong boxes and shipped by express. The less valuable roots can be shipped in burlap sacks, boxes, barrels, etc., by freight.

The various roots, barks, leaves, plants, etc., as described in this book are found thruout America. Of course there is no state where all grow wild, but there are many sections where several do. After reading this book carefully you will no doubt be able to distinguish those of value.

CHAPTER XVI.

Plants are of three classes--annuals, biennials, perennials. Annuals grow from seed to maturity in one year and die; biennials do not flower or produce seed the first year, but do the second and die; perennials are plants which live more than two years. Ginseng plants are perennial.

Roots, leaves, barks, etc., should be spread out thin in some dry, shady place. A barn floor or loft in some shed is a good place, providing it is light and "airy," altho the direct sunlight should not shine upon the articles being "cured." Watch while curing and turn or stir each day.

Prices given for roots, plants, leaves, etc., were those paid by dealers during 1907 unless otherwise specified. These prices, of course, were paid in the leading markets for fair sized lots. If you have only a few pounds or sold at some local market the price received was probably much less. The demand for the various articles varies and, of course, this influences prices--when an article is in demand prices are best.

After studying the "habitat and range" of the various plants as published together with the illustrations, there should be no difficulty in determining the various plants. By "habitat" is meant the natural abode, character of soil, etc., in which the plant thrives best and is found growing wild. To illustrate: Seneca Snakeroot--habitat and range--rocky woods and hillsides are its favorite haunts. It is found in such places from New Brunswick, Canada and Western New England States to Minnesota and the Canadian Rocky Mountains, and south along the Allegheny Mountains to North Carolina and Missouri.

From this it will be seen that it is useless to look for this plant in the Southern States, on the plains or in old cultivated fields, for such places are not its natural home.

CHAPTER XVII.

GOLDEN SEAL CULTIVATION.

I learned when a boy, by actual experience, that Golden Seal and Ginseng will not grow in open cultivated fields or gardens. I tried it faithfully. The soil must be virgin, or made practically so by the application of actual "new land" in such quantities that to prepare an acre for the proper growth of these plants would be almost impossible. And to furnish and keep in repair artificial shade for, say, an acre, would cost quite a little fortune. Of course one may cultivate a few hundred or few thousand in artificially prepared beds and shaded by artificial means, but to raise these plants successfully in anything like large quantities we must let nature herself prepare the beds and the shade.

When we follow nature closely we will not be troubled with diseases, such as blight and fungus. I know this by actual experience dear, and therefore dear to me.

Plants propagate themselves naturally by seedage, root suckers, and by root formation upon the tips of pendulous boughs coming in contact with the ground. Man propagates them artificially in various ways, as by layering, cuttings, grafting or budding, in all of which he must follow nature. The Golden Seal plant is readily propagated by any of the three following methods: (1) by seed; (2) by division of the large roots; (3) by suckers, or small plants which form on the large fibrous roots.

The seed berries should be gathered as soon as ripe, and mashed into a pulp, and left alone a day or two in a vessel, then washed out carefully and the seed stored in boxes of sandy loam on layers of rock moss, the moss turned bottom side up and the seed scattered thickly over it, then cover with about one-half inch of sandy loam, then place another layer of moss and seed, until you have four or five layers in a box. The box may be of any convenient size. The bottom of the box should be perforated with auger holes to secure good drainage. If water be allowed to stand upon the seeds they will not germinate, neither will they germinate if they become dry. The seeds should be kept moist but not wet. They may be sown in the fall, but, I think the

CHAPTER XVII.

better way, by far, is to keep your box of seeds in a cellar where they will not freeze until the latter part of winter or very early spring. If your seeds have been properly stratified and properly kept you will find by the middle of January that each little black seed has burst open and is wearing a beautiful shining golden vest. In fact, it is beginning to germinate, and the sooner it is put into the seed-bed the better. If left too long in the box you will find, to your displeasure, a mass of tangled golden thread-like rootlets and leaflets, a total loss.

To prepare a seed-bed, simply rake off the forest leaves from a spot of ground where the soil is rich and loamy, then with your rake make a shallow bed, scatter the seeds over it, broadcast, being careful not to sow them too thick. Firm the earth upon them with the back of the hoe or tramp them with the feet. This bed should not be near a large tree of any kind, and should be protected from the sun, especially from noon to 3 P. M.

The Golden Seal seedling has two round seed leaves upon long stems during the first season of its growth. These seed leaves do not resemble the leaves of the Golden Seal plant. The second and usually the third years the plant has one leaf. These seedlings may be set in rows in beds for cultivation in the early spring of the second or third year. This plant grows very slowly from seed for the first two or three years, after which the growth is more satisfactory.

By the second method, i. e., by division of the large roots, simply cut the roots up into pieces about one-fourth inch long and stratify in the same way as recommended for seeds, and by spring each piece will have developed a bud, and will be ready to transplant into beds for cultivation. This is a very satisfactory and a very successful method of propagating this plant. The plants grow off strong and robust from the start and soon become seed bearing.

[Illustration: Young Golden Seal Plant in Bloom.]

By the third method we simply let nature do the work. If the plants are growing in rich, loose, loamy soil, so the fibrous roots may easily run in every direction, the whole bed will soon be thickly set with plants. These may be taken up and transplanted or may be allowed to grow

CHAPTER XVII.

and develop where they are.

This is the method by which I propagate nearly all of my plants. It is a natural way and the easiest of the three ways to practice.

As to the proper soil and location for a Golden Seal garden I would recommend a northern or northeastern exposure. The soil should be well drained and capable of a thrifty growth of deciduous trees. It should contain an ample supply of humus made of leaf mold. It will then be naturally loose and adapted to the growth of Golden Seal. Cut out all undergrowth and leave for shade trees that will grow into value. I am growing locust trees for posts in my Golden Seal garden. I do not think fruit trees of any kind suitable for this purpose.

In preparing the ground for planting simply dig a trench with a mattock where you intend to set a row. This loosens up the soil and makes the setting easy. Set the plants in this row four to six inches apart. For convenience I make the rows up and down the hill. In setting spread the fibrous roots out each way from the large main root and cover with loose soil about one to two inches deep, firming the soil around the plant with the hands. Be very careful not to put the fibrous roots in a wad down in a hole. They do not grow that way. Plants may be set any time through the summer, spring or fall, if the weather he not too dry. The tops will sometimes die down, in which case the root will generally send up a new top in a few days. If it does not it will form a bud and prepare for growth the next spring. The root seldom if ever dies from transplanting. I know of no plant that is surer to grow when transplanted than Golden Seal. I make the rows one foot to fifteen inches apart. It does not matter as it will soon fill the spaces with sucker plants any way.

The cultivation of Golden Seal is very simple. If you have a deep, loose soil filled with the necessary humus your work will be to rid the plot of weeds, and each fall add to the fall of forest leaves a mulch of rotten leaves.

Do not set the plants deeper than they grew in a natural state, say about 1/2 to 3/4 inch. Spread the fibrous roots out in all directions and cover with leaf mold or some fine, loamy new soil. Water if the ground

CHAPTER XVII.

be at all dry. Then mulch with old forest leaves that have begun to decay. Let the mulch be about three or four inches deep and held on by a few light brush. The wind would blow the leaves away if not thus held in place. Be careful, however, not to press the leaves down with weights.

[Illustration: Golden Seal Plants.]

Remove the brush in the early spring, but let the leaves remain. The plants will come up thru them all right. This plant grows best in a soil made up entirely of decayed vegetation, such as old leaf beds and where old logs have rotted and fallen back to earth. If weeds or grass begin to grow in your beds pull them up before they get a start. Be careful to do this. Do not hoe or dig up the soil any way, The fibrous roots spread out in all directions just under the mulch. To dig this up would very much injure the plants.

I think the plants should be set in rows about one foot apart, and the plants three or four inches apart in the rows. This would require about 1,000 plants to set one square rod. My Golden Seal garden is in a grove of young locust trees that are rapidly growing into posts and cash. The leaves drop down upon my Golden Seal and mulch it sufficiently. The locust belongs to the Leguminous family of plants, so while the leaves furnish the necessary shade they drink in the nitrogen from the atmosphere and deposit great stores of it in the soil. This makes the soil porous and loose and gives the plant a very healthy dark green appearance.

We have only to follow the natural manner of the growth of the Golden Seal to be successful in its culture. Select a piece of sloping land, so as to be well drained, on the north or northeast side of the hill--virgin soil if possible. Let the soil be rich and loamy, full of leaf mold and covered with rotting leaves and vegetation. This is the sort of soil that Golden Seal grows in, naturally.

It is hard to fix up a piece of ground, artificially, as nature prepares it, for a wild plant to grow in. So select a piece if possible that nature has prepared for you. Do not clear your land. Only cut away the larger timber. Leave the smaller stuff to grow and shade your plants. There is no shade that will equal a natural one for Ginseng or Golden Seal.

Now, take a garden line and stretch it up and down the hill the distance you want your bed to be wide. Mark the place for the row along the line with a mattock, and dig up the soil to loosen it, so as to set the plants, or, rather, plant the roots easily. With a garden dibble, or some other like tool make a place for each plant. Set the plants 4 to 6 inches apart in the row. The crown of the plant or bud should be set about 1 inch beneath the surface.

Firm the earth around the plant carefully. This is an important point and should be observed in setting any plant. More plants are lost each year by carelessly leaving the earth loose over and around the roots than from any other cause. Do not leave a trench in the row. This may start a wash. Let the rows be about 1 foot apart. If land is no item to you, the rows may be further apart. They will, if properly cared for, in a few years, by sending up sprouts from the roots, fill up the end completely.

[Illustration: Thrifty Golden Seal Plant.]

When you have finished setting your bed, cover it with a good mulch of rotten leaves from the forest and throw upon them some brush to keep the wind from blowing them away. By spring the leaves will settle down compactly and you will be pleased to see your plants grow luxuriously. October and November are the best months of the year in which to set Golden Seal plants. They are, also, the months in which it should be dug for the market. It may be set in the spring if the plants are near by. The roots will always grow if not allowed to dry before transplanting.

If your bed does not supply you with plants fast enough by suckering, you may propagate plants by cutting the roots into pieces about one-fourth inch long, leaving as many fibrous roots on each piece as possible. These cuttings should be made in September or October and placed in boxes of sand over winter. The boxes should be kept in a cellar where they will not freeze. By spring these pieces will have developed a bud and be ready for transplanting, which should be done just as early as the frost leaves the ground so it can be worked.

CHAPTER XVII.

All the culture needed by this plant is to mulch the beds with forest leaves each fall and keep it clear of grass and field weeds. Wild weeds do not seem to injure it.

Golden Seal transplants easily and responds readily to proper cultivation. There is no witchcraft in it. The seeds ripen in a large red berry in July to germinate, if planted at once, the next spring. The fibrous roots, if stratified in sand loam in the autumn, will produce fine plants. Any good, fresh, loamy soil, that is partially shaded will produce a good Golden Seal.

You want soil that is in good tilth, full of humus and life, and free from grasses and weeds. It will stand a great deal more sunlight than Ginseng. It will also produce a crop of marketable roots much quicker than Ginseng. There is no danger of an over supplied market, as the whims of a nation changing, or of a boycott of a jealous people. I have my little patch of Golden Seal that I am watching and with which I am experimenting.

I want to say right here that you do not need a large capital to begin the culture of these plants that are today being exploited by different parties for cultivation. Just get a little plot of virgin soil, say six yards long by one yard wide and divide it into two equal lots. Then secure from the woods or from some one who has stock to sell about 100 plants of each, then cultivate or care for your apron garden and increase your plantation from your beds as you increase in wisdom and in the knowledge of the culture of these plants.

The Bible says "Despise not the day of small things." Do not, for your own sake, invest a lot of money in a "Seng" or Seal plantation or take stock in any exploiter's scheme to get rich quick by the culture of these plants. Some one has written a book entitled "Farming by Inches." It is a good book and should be in every gardener's library. Now, if there be any crops that will pay a big dividend on the investment farmed by inches "Seng" and Seal are the crops.

CHAPTER XVIII.

CHAPTER XVIII.

GOLDEN SEAL, HISTORY, ETC.

The increasing use of Golden Seal in medicine has resulted in a wide demand for information about the plant, its identification, geographical distribution, the conditions under which it grows, methods of collecting and preparing the rhizomes, relations of supply and demand, and the possibilities of its cultivation. This paper with the exception of the part relating to cultivation was prepared (under the direction of Dr. Rodney H. True, Physiologist in Charge of Drug and Medicinal Plant Investigations) by Miss Alice Henkel, Assistant in Drug and Medicinal Plant Investigations; and Mr. G. Fred Klugh, Scientific Assistant in the same office, in charge of Cultural Experiments in the Testing Gardens, furnished the part treating of the cultivation of this plant. In the preparation of this paper, which was undertaken to meet the demand for information relative to Golden Seal, now fast disappearing from our forests, many facts have been obtained from Lloyd's Drugs and Medicines of North America.

Lyster H. Dewey, Acting Botanist. Office of Botanical Investigations and Experiments, Washington, D. C, Sept. 7, 1904.

History.

As in the case of many other native medicinal plants, the early settlers learned of the virtues of Golden Seal thru the American Indians, who used the root as a medicine and the yellow juice as a stain for their faces and a dye for their clothing.

The Indians regarded Golden Seal as a specific for sore and inflamed eyes and it was a very popular remedy with pioneers of Ohio and Kentucky for this affliction, as also for sore mouth, the root being chewed for the relief of the last named trouble.

Barton in his "Collection for an Essay towards a Materia Medica of the United States," 1804, speaks of the use of a spiritual infusion of the root of Golden Seal as a tonic bitters in the western part of Pennsylvania and the employment of an infusion of the root in cold

CHAPTER XVIII.   128

water as a wash for inflammation of the eyes.

According to Dr. C. S. Rafinesque, in his Medical Flora in 1829, the Indians also employed the juice or infusion for many "external complaints, as a topic tonic" and that "some Indians employ it as a diuretic stimulant and escharotic, using the powder for blistering and the infusion for the dropsy."

He states further that "internally it is used as a bitter tonic, in infusion or tincture, in disorders of the stomach, the liver," etc.

It was not until the demand was created for Golden Seal by the eclectic school of practitioners, about 1747, that it became an article of commerce, and in 1860 the root was made official in the Pharmacopoeia of the United States, which place it has held to the present time.

Habitat and Range.

Golden Seal occurs in patches in high open woods where there is plenty of leaf mold, and usually on hillsides or bluffs affording nature drainage, but it is not found in very moist or swampy situations, in prairie land, or in sterile soil. It is native from southern New York to Minnesota and western Ontario, south to Georgia and Missouri, ascending to an altitude of 2,500 feet in Virginia. It is now becoming scarce thruout its range. Not all of this region, however, produced Golden Seal in abundance. Ohio, Indiana, Kentucky and West Virginia have been the greatest Golden Seal producing states, while in some localities in southern Illinois, southern Missouri, northern Arkansas, and central and western Tennessee the plant, tho common, could not be said to be sufficiently plentiful to furnish any large amount of the root. In other portions of its range it is sparingly distributed.

Common Names.

Many common names have been applied to this plant in different localities, most of them bearing some reference to the characteristic yellow color of the root, such as yellow root, yellow puccoon, orange-root, yellow paint, yellow Indian paint, golden root, Indian dye,

CHAPTER XVIII.

curcuma, wild curcuma, wild tumeric, Indian tumeric, jaundice root and yellow eye; other names are eyebalm, eye-root and ground raspberry. Yellow root, a popular name for it, is misleading, as it has been applied to other plants also, namely to gold thread, false bittersweet, twinleaf and the yellow-wood. The name Golden Seal, derived from its yellow color and seal-like scars on the root, has been, however, generally adopted.

Description of the Plant.

It is a perennial plant and the thick yellow rootstock sends up an erect, hairy stem about a foot in height, around the base of which are two or three yellowish scales. The stems, as they emerge from the ground, are bent over, the tops still remaining underground, and sometimes the stems show some distance above the surface before the tops are brought out from the soil. The yellow color of the roots and scales extends partly up the stem so far as it is covered by soil, while the portion of the stem above the ground has a purplish color. Golden seal has only two leaves (rarely three), the stem bearing these seeming to fork at the top, one branch supporting a large leaf and the other a smaller one and a flower. Occasionally there is a third leaf, much smaller than the other two and stemless.

The leaves are prominently veined on the lower surface, and are palmately 5 to 9 lobed, the lobes broad, acute, sharply and unequally toothed. The leaves are only partially developed at flowering time and are very much wrinkled, but they continue to expand until they are from 6 to 8 inches in diameter, becoming thinner in texture and smoother. The upper leaf subtends or encloses the flower bud.

Early in spring, about April or May, the flower appears, but few ever see it as it lasts only five or six days. It is greenish-white, less than half an inch in diameter, and has no petals, but instead three small petal-like sepals, which fall away as soon as the flower expands, leaving only the stamens--as many as 40 or 50--in the center of which are about a dozen pistils, which finally develop into a round, fleshy, berrylike head. The fruit ripens in July or August, turning a bright red and resembling a large raspberry, whence the common name ground raspberry, is derived. Each fruit contains from 10 to 20 small, black, shining, hard seeds.

If the season has been moist, the plant sometimes persists to the beginning of winter, but if it has been a dry season it dies soon after the fruit is ripe, so that by the end of September no trace of the plant remains above the ground. In a patch of Golden Seal there are always many sterile stems, simple and erect, bearing a solitary leaf at the apex but no flower.

Mr. Homer Bowers, of Montgomery county, Ind., who propagated Golden Seal from the seed for the purpose of studying its germination and growth, states that the plant grown from naturally sown seed often escapes observation during the first year of its existence owing to the fact that in this entire period nothing but two round seed leaves are produced and at this stage the plant does not look materially different from other young seedings. During its second year from seed one basal leaf is sent up, followed in the third year by another smaller leaf and the flower.

Description of the Rhizome, or Rootstock.

The rhizome (rootstock) and rootlets of Golden Seal, or hydrastis, as it is also known in the drug trade, are the parts employed in medicine. The full-grown rhizome, when fresh, is of a bright yellow color, both internally and externally, about 1 1/2 to 2 1/2 inches in length, and from one-fourth to three-fourths of an inch in thickness. Fibrous yellow rootlets are produced from the sides of the rhizome. The fresh rhizome contains a large amount of yellow juice, and gives off a rank, nauseating odor. When dry the rhizome measures from one to two inches in length and from one-eighth to one-third of an inch in diameter.

It is crooked, knotty, wrinkled, of a dull brown color outside, and breaks with a clean, short, resinous fracture, showing a lemon-yellow color if the root is not old. If the dried root is kept for a long time it will be greenish-yellow or brown internally, and becomes inferior in quality. On the upper surface of the rhizome are several depressions, left by former annual stems, which resemble the imprint of a seal; hence the name Golden Seal.

The fibrous rootlets become very wiry and brittle in drying, break off readily and leaving only small protuberances, so that the root as found in commerce is sometimes almost bare. The dried rhizome has also a peculiar, somewhat narcotic, disagreeable odor, but not so pronounced as in the fresh material; an exceedingly bitter taste; and a persistent acridity which causes an abundant flow of saliva when the rhizome is chewed.

Collection and Preparation of the Root.

The root should be collected in autumn after the plants have matured. Spring-dug root shrinks far more in drying and always commands a lower price than the fall-dug root. After the roots are removed from the earth they should be carefully freed from soil and all foreign particles. They should then be sorted and small, undeveloped roots and broken pieces may be laid aside for replanting. After the roots have been cleaned and sorted they are ready to be dried or cured.

Great care and judgment are necessary in drying the roots. It is absolutely necessary that they should be perfectly dry before packing and storing, as the presence of moisture induces the development of molds and mildews, and of course renders them worthless. The roots are dried by the exposure to the air, being spread out in thin layers on drying frames or upon a large, clean, dry floor. They should be turned several times during the day, repeating this day after day until the roots are thoroughly dried. If dried out of doors they should be placed under cover upon indication of rain and at night so that they may not be injured by dew. After the roots are thoroughly dried they may be packed as tightly as possible in dry sacks or barrels and they are then ready for shipment.

Diminution of Supply.

Altho, perhaps, in some secluded localities Golden Seal may still be found rather abundantly, the supply is rapidly diminishing and there is a growing scarcity of the plant thruout its range. With the advance of civilization and increase in population came a growing demand for many of our native medicinal plants and a corresponding decrease in the sources of supply. As the rich forest lands of the Ohio valley and

CHAPTER XVIII.

elsewhere were required for the needs of the early settlers they were cleared of timber and cultivated, and the Golden Seal, deprived of the shelter and protection necessary to its existence, gradually disappeared, as it will not thrive on land that is cultivated.

Where it was not destroyed in this manner the root diggers, diligently plying their vocation, did their share toward exterminating this useful little plant, which they collected regardless of the season, either before the plants had made much growth in the spring or before the seeds had matured and been disseminated, thus destroying all means of propagation. The demand for the root appears to be increasing, and the time seems to be not far distant when this plant will have become practically exterminated, so far as the drug supply is concerned.

The cultivation of golden seal seems now to have become a necessity in order to meet the demand and save the plant from extinction. Prior to 1900 there seemed to be no one, so far as the Department of Agriculture could ascertain, who had ever attempted the cultivation of golden seal for the market. From that time on, many inquiries were directed to the Department by persons who were quick to note the upward tendency of prices for golden seal and there are now several growers in different parts of the country who have undertaken the cultivation of golden seal on a commercial scale.

Cultivation.

The United States Department of Agriculture has been carrying on experiments in the cultivation of Golden Seal on a small scale at Washington, D. C., since the spring of 1899, in the hope that methods might be worked out according to which this valuable wild drug plant could be grown on a commercial scale. In these experiments the aim has been to imitate the natural conditions of growth as closely as possible. The results that have thus far been obtained, while not as complete in some respects as would be desirable, seem to justify the conclusion that Golden Seal can be successfully cultivated. The methods of operation described apply to the conditions at Washington, and the treatment may need to be somewhat modified under other conditions of soil and climate.

CHAPTER XVIII.

Necessary Soil Conditions.

The soil conditions should imitate as closely as possible those seen in thrifty deciduous forest. The soil should contain an ample supply of humus, well worked into the ground, to secure the lightness and moisture-retaining property of forest soils. The best form of humus is probably leaf mold, but good results may be obtained by mulching in the autumn or early winter with leaves, straw, stable manure, or similar materials.

After the soil has been prepared and planted, it is well to add a mulch in the fall as a partial protection to the roots during the winter, and the decay of this material adds to the value of the soil by the time the plants appear in the spring. The forest conditions are thus imitated by the annual addition of vegetable matter to the soil, which by its gradual decay accumulates an increasing depth of a soil rich in materials adapted to the feeding of the plants and to the preservation of proper physical conditions.

The growth of the weeds is also hindered to a considerable extent. If sufficient attention is given to the presence of this mulch, the nature of the underlying soil is of less importance than otherwise. In the case of clay the thorough incorporation of a large amount of decayed vegetable matter tends to give lightness to the otherwise heavy soil, facilitating aeration and drainage. Since the roots of the Golden Seal do not grow well in a wet soil, thorough drainage is necessary. A lighter, sandy soil is improved by the addition of humus, since its capacity to hold moisture is thereby increased and the degree of fertility is improved.

The looser the soil, the easier it is to remove the roots in digging without breaking or injuring them. Before planting, the soil should be thoroughly prepared to a depth of at least 6 or 8 inches, so as to secure good aeration and drainage. The good tilth thus secured will be in a degree preserved by the continued addition of the mulch. A further advantage of a careful preparation is seen in a decrease in the amount of cultivation required later.

Artificial Shade.

Since the Golden Seal grows naturally in the woods, it must be protected from the full light of the sun by artificial shade. That used in connection with the experiments of the Department was made of ordinary pine plastering lath, nailed to a suitable frame elevated on posts. The posts were of cedar 8 1/2 feet long, set 2 1/2 feet in the ground in rows 11 feet apart, and 16 feet distance from each other in the rows. Supports 2 by 4 inches were set on cedar blocks 2 feet long sunk below the soil surface in the middle of the 16-foot spaces. Pine pieces 2 by 4 inches were nailed edgewise to the tops of the posts and supports. The posts were notched to receive the 2 by 4-inch sticks. Pieces 2 by 4 inches were nailed across these at intervals of 4 feet. The laths were nailed to these, leaving spaces about an inch wide.

This shade has been found to be satisfactory, as it is high enough above the ground to allow such work as is necessary in preparing and cultivating the land. If the lathing is extended 2 or 3 feet beyond the posts on the sunny sides, injury from the sun's rays at the edges of the area will be prevented. The sides may be protected by portable board walls about 2 feet high set around the edges. Protection from injury by winds when the tops are large may be thus secured. Too much dampness should be guarded against in the use of the board sides, since conditions might be developed favorable to the damping off fungus and to aphides during the hot, rainy periods.

Trees may be used for shade, but this is in some ways to be regarded as unsatisfactory. When the shade produced is of the right density, the use of the moisture and raw food materials of the soil by the trees is an undesirable feature.

Attention Required.

The cultivation of Golden Seal is simple. Having secured a deep, loose soil, rich in humus, renewed annually by the application of a new mulch, the removal of weeds is the chief care. The soil, if properly prepared, will tend to maintain itself in good condition. The manner of treatment is very similar to that required by Ginseng, which is also a plant of moist woods. If the ground is thoroughly prepared, beds are not absolutely necessary. The plants may be grown in rows 1 foot apart and 6 inches apart in the rows. Beds may be thought by some to

CHAPTER XVIII.   135

be more convenient, enabling the grower to remove the weeds and collect the seed more readily. If beds are used, they may be made from 4 to 8 feet wide, running the entire length of the shade, with walks from 18 inches to 2 feet wide between. Boards 6 or 8 inches wide are set up around the sides of the beds, being held in place by stakes driven on each side of the board in the center and at the ends. These beds are filled with prepared soil, and the plants are set 8 inches apart each way.

Methods of Propagation.

There are three possible ways of propagating the plant: (1) by seed; (2) by division of the rhizomes; (3) by means of small plants formed on the stronger fibrous roots. Thus far no success has been attained in growing Golden Seal from the seed. The second and third methods have given better results.

Experiments With Seeds.

Seeds just after ripening were planted in sandy soil mixed with well rotted stable manure and mulched lightly with manure. Other lots were kept over winter in a dry condition and planted in the spring in potting soil in a greenhouse. No seedlings have appeared, but a long rest period may be demanded and the seed may yet germinate.

Experiments With Divided Rhizomes.

In the spring of 1902, 40 plants were secured and planted under a shade of temporary character, but the season was too far advanced to permit of much growth during that year. In 1903, proper shade was supplied, all other conditions were better, and the plants made a good growth. The crop was dug about the middle of November 1903; the roots were weighed and divided. They were again planted and in May, 1904, there were found to be 150 strong plants and a few smaller ones as a result of this division, an increase of 275 per cent.

This method of propagation seems to be the most important and the other two of second importance. The processes are simple and no skill is needed. The plant dies down in late summer and the stem decays, leaving a scar in its place on the rhizome. Two or more buds are formed on the sides of the rhizome and these accumulate energy for growth the following spring. If the root is cut in as many pieces as there are buds, giving each plant a portion of the rhizome, some fibrous roots, and one or more buds, the number of the plants can be doubled. The roots are planted and mulched and the process is complete. The rains pack the soil around the roots and they are ready to grow when spring comes. The process may be repeated every year and the number of roots increased indefinitely.

The stronger fibrous roots of the larger plants dug in the autumn of 1903 were formed from a few inches to a foot from the rhizome. Some were about half an inch long, but the majority of them were smaller. The larger ones need no special treatment and may be planted with the main crop. The smaller ones should be planted in boxes or beds of well prepared soil, at a distance of about 3 inches apart, mulched with a thin coating of leaf mold or similar material, and grown in shade until large enough to transplant to the shelter with the larger plants. They will probably require at least three years to reach their full development.

If they could be left undisturbed in the beds where they are formed they would receive nourishment from the older rhizomes and perhaps grow faster, but it is probably best to divide the older roots every year where propagation alone is desired, planting the smaller roots and the plants made by division of the rhizomes. The larger roots are marketed to more advantage than the smaller ones, so it is best to have the surplus consist of the larger roots. The frequent working of the soil allowed by this treatment will keep it in better condition than if left undisturbed for a longer period.

Yield of Roots.

The yield from the small plant grown by the Department was 4 pounds of green roots to an eighth of a square rod of soil, or 5,120 pounds per acre. This, when dried, would give about 1,500 pounds of marketable roots. The conditions were not very good, the shade being too close to the plants and the plants being set too far apart. The yield will probably be larger with the shade now in use. The 150 roots obtained by

CHAPTER XVIII. 137

dividing the above crop now occupy less than one-fourth of a square rod and are set in rows one foot apart and 6 inches apart in the rows.

Time Necessary to Mature Crop.

The number of years necessary to produce the largest crop has not been definitely determined, but the roots begin to decay after the fourth year and the central and largest part of the root decays at the oldest scar, leaving two or more plants in place of the old one. No advantage can be gained by growing the plants more than three years and probably very little by growing them more than two years. For propagation alone, one year will give good results, while for maintaining a constant area and producing a crop, two or three years, depending upon the growth made, will give a good crop of large, marketable roots.

Market Conditions.

Golden Seal is a root the price of which has fluctuated widely, because of the alternate oversupply and scarcity, manipulation of the market, lack of demand, or other influences. High prices will cause the diggers to gather the root in abundance, thus overstocking the market, which the next season results in lower prices, at which diggers refuse to collect the root, thus again causing a shortage in the supply. Lack of demand usually brings about a shrinkage in price, even tho the supply is light, while an active demand will cause prices to advance in spite of a plentiful supply.

The arrival of spring dug root has a weakening effect on the market, altho the fall dug root is always preferred. For the past few years, however, high prices have been steadily maintained and there appears to be but one cause for this and that is, as already pointed out, that the forests no longer yield unlimited quantities of this valuable root, as in former years, and the scant supply that can be had is inadequate to meet the constantly increasing demand.

According to the market reports contained in the Oil, Paint and Drug Reporter, the year 1904 opened with a quotation of 74 to 75 cents, will soon advance (in one week early in February) from 76 cents to 95 cents. A still further advance occurred about the end of February, when the price went up from $1.00 to $1.25 per pound. In March the market was almost destitute of supplies, but lack of interest brought the price down to $1.10. In May the price again advanced to $1.25 and it was stated that the local supplies were being held by a small number of dealers, altho it was believed that together they held not more than 1,000 pounds. About June 1st the arrival of spring dug roots caused the market to sag, prices ranging from $1.10 to $1.18 during that month and in July from 90 cents to $1.10.

In August the lowest price was $1.15 and the highest $1.50, no discrimination being made between the fall dug and the spring dug roots. From September 1st to October 15th, 1904, the price of Golden Seal varied but little, $1.35 being the lowest and $1.40 the highest quotation. No supplies worth mentioning can be obtained in the West; the stock in New York is short and the demand, especially for export, is increasing. It is impossible to ascertain the exact annual consumption of Golden Seal root, but the estimates furnished by reliable dealers place these figures at from 200,000 to 300,000 pounds annually, about one-tenth of which is probably used for export.

It will be observed that the price of this article is very sensitive to market conditions and it seems probable that the point of overproduction would be easily reached if a large number of Golden Seal growers were to meet with success in growing large areas of this drug.

By Alice Henkel, Assistant, and G. Fred Flugh, Scientific Assistant, Drug and Medicinal Plant Investigations. U. S. Department of Agriculture.

CHAPTER XIX.

GROWERS' LETTERS.

Considerable has been said the past few years concerning Hydrastis (Golden Seal) and I do not wish to enter on a long article describing this plant, but will make the facts brief and narrate some of my experiences with the plant under cultivation.

The scientific name is Hydrastis Candensis, the common name Golden Seal, yellow root, puccoon root, Indian tumeric, etc., according to the section in which it is found. It is a perennial plant with an annual stem same as Ginseng, and appears above ground in the spring at the same time and manner. The stalk coming thru the ground bent and leaves folded. It has from one to three palmately five to nine lobed leaves, uneven and sharply toothed.

The fruit or seed grows from the base of one of these leaves. Flower is first whitish green producing the fruit red and resembling a strawberry, maturing last of July and the first of August.

The berry contains from fifteen to twenty small oval black shinny seeds. Only a portion of the stalks ever bear seed. From the middle to the last of September the stalks die down and when winter comes on the hydrastis bed appears the same as a Ginseng bed.

The root stalk or rhizome is thick, rough covered with rounded indentations or eyes, dark yellow in color and having many long threadlike bright yellow fibres branching in all directions. It has one and sometimes as many as four buds which will produce the next season's stalks. Besides these there are many latent buds and little plantlets on the runners of fibrous roots.

The root and all of its fibres is the part used in medicine.

I presume it will be difficult to fix a date when this plant was first used in medicine. But it is known that the Indians used it in healing diseases and in preparing stains and paints when first observed by the white man. Dr. Rafinesque first makes mention of it in a medical work in

CHAPTER XIX.

1828 and the elective physicians adopted it in their practice in 1847. The Pharmacopoeia of the U. S. in 1860 made Hydrastis an official drug and described the manufacture of different preparations.

It has since gained in favor and in extent of application until at present it is almost the specific in the treatment of certain catarrhal conditions. Thousands of pounds being used by the physicians in different parts of the world variously estimated from 200,000 to 300,000 pounds annually, more extensively, as you see, than Ginseng.

The price has advanced as given by the Drug Reporter, from 1894 of 18 to 23 cents a pound, to 1903, of 52 to 75 cents a pound, since 1903 to 1906 it has advanced to $1.10 to $1.30 a pound. The figures representing the highest and lowest quotations of those years. The price of the plant has advanced first because investigation has proven the value of the plant as a drug in the healing art increasing its consumption, second the consumption of and destruction of its habitat is limiting its supply. It is used in all countries, but not found in all countries in its wild state. The United States supplies the majority of the root.

Its cultivation is very promising and profitable because only very few have entered the industry yet, the wild supply is becoming exhausted, the drug trade demands it and its consumption depends upon a sound demand.

There is a promising opportunity in this industry and when I am speaking I am not offering inducements to get the rich quick individual, but to the careful, painstaking, plodding individual who is willing to give at least some labor for a handsome compensation. I have been one of the pioneers to begin the investigation and cultivation of this plant, and shall tell some of my experience in handling the plants.

[Illustration: Golden Seal in an Upland Grove.]

I procured four years ago several pounds of green Hydrastis root from a digger and set them out in three different patches. One in the open garden, one in an inclosure shaded in the garden, and one bed in a grove. I had the beds made the same as instructions had been given

CHAPTER XIX.

me for making beds for Ginseng. Ground loose and mellow, I selected only roots with buds formed, and set an inch under ground and six inches apart.

This was in June. All the plants came up and all made a good growth except those in the open, the leaves on these remained small and pinched about two to three inches from the ground. In digging them I found that they had thrown out a number of fibrous roots. In the fall I procured and set several thousand roots in the woods.

The next fall I set many more, but this time I cut the roots into three or four pieces and planted. All came the next summer, some not appearing above ground until June. I have had no success in planting seeds, so do not use this means of raising the plants. The method I use now is to cut the roots across so a latent bud will be on one piece, all small pieces broken and the fibers for some of these grow a plant.

After preparing the beds loose I lay little trenches across and drop the pieces in these every two or three inches apart, then cover about an inch with loose dirt, then leaves and mulch. The best time I have found to plant is in September, the earlier the better, for the buds then form before freezing up and are ready to come in the spring early.

They grow larger and thriftier if well rotted manure is in the ground and this does not interfere with the quality of the root. The largest roots I have seen grew in a hog lot supplied with hog manure. In three or four years I dig the roots, using a manure fork, the largest ones I wash and dry; the smaller ones and pieces I use for planting.

I am arranging a barrel shaped affair closed at the ends and covered around with wire to wash the roots. The method is to put a rod thru with handles on ends and rest on grooves on posts immersed half way of barrel in running water and revolve. In this way I believe the roots can be washed readily by splashing and falling in the water, and tons of the roots easily handled and washed clean with little help.

I have dried them by spreading on racks to dry in the sun. In bright sun it requires two or three days. As they wilt, I place on paper in order to save the fibres that break off. When making a business of growing

CHAPTER XIX.

these roots and having good, fresh roots in considerable quantity, a better price can be commanded by dealing direct with the drug mill. A great many of the roots when dug will weigh one ounce or more and the roots lose in weight about the same as drying Ginseng.

Dr. L. C. Ingram, Wabasha County, Minn.

* * *

There has never been a time in the history of this country when the cultivation of certain medicinal plants, as Golden Seal, Ginseng, Seneca and others appealed so much to those interested in such things as the present.

Many of these plants have hitherto been found growing wild in our woods and fields, and along our road sides and waste places, and have usually been gathered in an immature state and out of season, washed and cured in a slovenly manner and bartered at country stores for coffee and calico and other commodities. In this way the drugs and drug trade of the country have been supplied. I think it is very evident to the casual observer that this manner of supply is nearing its close finally and forever.

The merchant who handles the stock may not know as yet the great and growing scarcity of almost all our medicinal plants. But the digger who has stood at the first end of the drug trade, in touch with the natural supply, knows that the fountains are dried up, in great measure, and that the streams of the trade must necessarily soon cease to flow or be supplied by artificial means. In most cases medicinal plants grow naturally in the best soils, the sandy, loamy, moist north hill sides, the rich, black coves at the heads of our small streams and in the rich alluvial bottoms along our larger creeks and small rivers. They will not grow in wet lands or on south hill sides. This should be remembered by the would-be culturist and the natural whims of the plant attended to, else failure and disappointment are sure.

What I have said is peculiarly the case with Golden Seal, the yellow root of our locality, the ground raspberry of another, the yellow puccoon of another and probably bearing other local names in other

CHAPTER XIX. 143

localities. The natural habitat of Golden Seal has been cleared up for farming or grazing purposes, while the keen eyed "sanger" has ferreted out every nook and corner adapted to the growth of this plant and then ruthlessly dug it, little and big, old and young, until today it is a very scarce article.

The Indians regarded Golden Seal as a sure remedy for sore and inflamed eyes, sore mouths, old sores, wounds, etc., and first taught the whites its use as a remedy.

The pioneers used it as teas, washes and salves years before it became known to the medical fraternity. It did not become an article of commerce in any way until about the year 1847, and then it was so plentiful and so little used that the trade was supplied at 3 cents per pound for the dried root. I dug it myself, when a boy, as late as 1868, and received 5 cents per pound for the dried root, in trade, at a country store. I found it plentiful in patches in open woods where the ground was rich and favored the growth of paw paw, dogwood, walnut, elm, sugar maple, etc. It grew best in land well drained and full of leaf mold. Remember this, ye planters.

Well, the demand has rapidly increased, and the supply, from the causes afore mentioned, has more rapidly decreased, until the price has risen from 3 cents to $1.50 per pound. Golden Seal was originally found growing in favorable localities from Southern New York west to Minnesota, thence south to Arkansas and east to Georgia and the hill regions of the Carolinas. Ohio, Indiana, West Virginia and Eastern Kentucky have been by far the greatest Golden Seal producing sections.

Golden Seal is a perennial plant, the gnarly, knotty root of which is the part used in medicine. These knotty roots send out in every direction many long, slender, bright yellow, fibrous roots. Each root in spring early sends up one to six hairy stems six inches to fifteen or twenty inches in height, each stem supporting at the top one, or if a seed yielding plant, two large leaves, in shape somewhat resembling the leaf of the sugar maple, but thicker and more leathery. At the base of each stem are two or three scale like leaves starting from the root, around the stem and extending to the surface of the ground. These

CHAPTER XIX.   144

scales are yellow while the leaf stems are somewhat purplish in color. The seed bearing stocks fork near the top of the plant, each stem supporting a leaf, the smaller leaf enclosing a flower bud at the base and at the top of the leaf stem. The plants that are not of seed bearing age and size do not fork and have but one leaf. The flowers are greenish, about an inch in diameter and open, here, about the first of May. Then continue open about five days when the petals fall and the development of the seed berry begins.

This berry ripens in July. When ripe it is red in color and resembles a large raspberry and contains about 20 to 30 small, round, black, shiny, hard seeds. These seeds, if stratified at once and kept in moist, sandy loam, will begin to open by the first of February, each seed showing a beautiful, bright, shiny, golden bud. The seeds should be planted very early. When it comes up the young plant has two leaves and does not develop any further leaf or stem growth during the first summer. The first two leaves do not look at all like those that follow. So, be careful or you will destroy your plants for weeds.

Plants may be readily propagated by cutting up the roots into pieces, say 1/4-inch long and placing these root cuttings in boxes of loamy sand in the autumn. By spring each root cutting will have developed a fine bud and be ready for transplanting, which should be done as early as possible. The plant also propagates itself by sending up suckers from the fibrous roots.

[Illustration: Locust Grove Seal Garden.]

As to culture, I would say, follow nature. Do not plow and hoe and rake and make a bed as for onions. Just simply select a piece of virgin soil, if possible, and make rows, say one foot apart and set the plants about three or four inches apart in the rows. All the culture needful is to pull out the weeds, and, if the trees in the patch be not sufficient to furnish a good leaf mulch in the fall, attend to this by mulching with a good coat of forest leaves.

My Golden Seal garden is in a locust grove that is rapidly growing into posts, so, you see, I am getting two very profitable crops off the same land at the same time. The plants should grow in a bed of this kind

CHAPTER XIX. 145

until it becomes full of roots, which will require three to five years. It is all the better if they are allowed to grow longer. The whole patch should be dug in the fall when the tops die down. The large roots should be carefully washed and cleansed of all foreign roots and fibers and dried on clean cloths in the shade, when it is ready for market and should be shipped in clean, new bags to some reliable dealer in the larger cities. There are plenty of them and I would advise that you write to several of them, telling them just what you have before you ship.

I know from actual experience that good money may be made by the right party in the culture of Golden Seal. If a young man would start a garden of medicinal plants and attend to it at odd times, studying the nature of the plants and carefully save all seeds and add them to his stock, in a few years he would have a garden with a large sum of money. I have estimated an acre of Golden Seal at full maturity and as thick on the ground as it should be grown to be worth $4,840, or one dollar per square yard. It will not take a very great while to fill an acre with plants. Besides, if the land is planted in locust trees it is yielding two crops of wonderful value at the same time.

One young man from Virginia says: "I have a piece of new ground just cleared up which I think would be just the thing, and then I could set out short stem red cherries to shade and cover the ground. Please let me hear from you at once." Well, if this piece of ground is on the right side of the hill, that is, the north or northeast or west slope, and is rich, loose and loamy, full of leaf mold and naturally well drained, it is all right for Golden Seal, but would it suit cherries? Cherries might do very well for shade, but I would prefer catalpa or locust or some other quick growing timber tree to any sort of fruit tree.

One reason is that in gathering the fruit and in caring for the trees I think the Golden Seal would be trampled upon and injured, also the ground would be trampled and compacted and thus rendered unsuitable for this plant. The ground in which Golden Seal grows should be kept in its "new state" as much as possible. However, my Virginia friend may succeed well with his cherries and Seal. He must keep up the primitive condition of the soil and keep out weeds and grass.

CHAPTER XIX.

Another question, "How long will it take it to mature?" As to its "maturity," it may be dug, cleansed, dried and marketed at any time and in any stage of its growth. But I think that a setting of Golden Seal should be dug in the fall three or four years after planting; the large roots washed and cleansed and made ready for market, while the smaller roots should be used for resetting the bed. You will have enough small roots to set a patch ten or twelve times the size of the one you dig, as each root set will in three or four years produce ten to fifteen good plants besides yielding a lot of seed.

"How much will it cost to plant one-eighth of an acre?" One-eighth of an acre contains twenty square rods, and to set one square rod, in rows eighteen inches apart would take 363 plants, and twenty square rods would take 20 times 363 plants, or 7,260 plants, which at $10.00 per thousand, would cost $72.60. But I would advise the beginner to "make haste slowly" in trying new things. A thing may be all right and very profitable if we understand it and give it proper culture, while it is very easy to make sad failure by over doing a good thing. So let the beginner procure a thousand or so plants and start his garden on a small scale, and increase his plantation from his own seed bed as his knowledge of the plant and its culture increases. A very large garden may be set in a few years from 1,000 plants.

"Should the seed be sown broadcast?" To be successful with the seed requires great patience and pains. I make a large flat brush heap and burn it off in the fall. I then dig up the ground to the depth of three or four inches and place boards edgewise around this bed, letting them down into the ground two or three inches. These boards are to keep out mice and to prevent washing. I then sow the seeds in little trenches made with a hoe handle about six inches apart and pretty thick in the trenches and smooth over and tramp solid.

Then sow a few handfuls of bone dust mulched with forest leaves and cover with brush to keep the leaves from blowing away. You are done now until spring. In the early spring, after freezing weather is over, carefully remove the brush and the mulch of leaves. Remember this must be done early as the plant wants to come up early. Watch for your young plants and carefully pull up every weed as soon as it shows itself. Mulch again in the fall and remove as before the next

CHAPTER XIX.

spring. Keep down weeds as before, and by fall you will have a fine lot of No. 1 two-year-old plants, which may be transplanted to the garden at once or early the next spring.

I should have stated that Golden Seal seed should not be allowed to dry after gathering. They should be placed in layers of sand in a box and kept moist until planting time. They begin to germinate very early, and if you delay planting until spring you are nearly sure to lose them.

As to the "profits," I want it distinctly understood that I do not think that every one who starts a bed or patch of Golden Seal will be a millionaire in a few years. But I do think, and in fact I know, that considering the land in cultivation, the time and expense of its culture, it is one of the most profitable crops that can be grown in this latitude.

Lee S. Dick, Wayne County, W. Va.

CHAPTER XX.

GOLDEN SEAL--GOVERNMENT DESCRIPTION, ETC.

The following is from a bulletin issued by the U. S. Department of Agriculture--Bureau of Plant Industry--and edited by Alice Henkel:

Hydrastis Canadensis L.

Pharmacopoeial Name--Hydrastis.

Other Common Names--Yellowroot, yellow puccoon, orange-root, yellow Indian-paint, turmeric-root, Indian turmeric, Ohio curcuma, ground raspberry, eye-root, eye-balm, yellow-eye, jaundice-root, Indian-dye.

Habitat and Range--This native forest plant occurs in patches in high, open woods, and usually on hill sides or bluffs affording natural drainage, from southern New York to Minnesota and western Ontario, south to Georgia and Missouri.

Golden Seal is now becoming scarce thruout its range. Ohio, Indiana, Kentucky and West Virginia have been the greatest Golden Seal producing states.

[Illustration: Golden Seal (Hydrastis Canadensis) Flowering Plant and Fruit.]

Description of Plant--Golden Seal is a perennial plant belonging to the same family as the buttercup, namely the crowfoot family (Ranunculaceae.) It has a thick yellow rootstock, which sends up an erect hairy stem about 1 foot in height, surrounded at the base by 2 or 3 yellowish scales. The yellow color of the roots and scales extends up the stem so far as it is covered by soil, while the portion of the stem above ground has a purplish color. The stem, which has only two leaves, seems to fork at the top, one branch bearing a large leaf and the other a smaller one and a flower. A third leaf, which is much smaller than the other two and stemless, is occasionally produced. The leaves are palmately 5 to 9 lobed, the lobes broad, acute, sharply and unequally toothed; they are prominently veined on the lower surface and at flowering time, when they are very much wrinkled, they are only partially developed, but they continue to expand until they are from 6 to 8 inches in diameter becoming thinner in texture and smoother. The upper leaf subtends or incloses the flower bud. The greenish white flower appears about April or May, but it is of short duration, lasting only five or six days. It is less than half an inch in diameter, and, instead of petals, has three small petal-like sepals, which fall away as soon as the flower expands, leaving only the numerous stamens (as many as 40 or 50), in the center of which are about a dozen pistils, which finally develop into a round fleshy, berry-like head which ripens in July or August. The fruit when ripe turns a bright red and resembles a large raspberry, whence the common name "ground-raspberry" is derived. It contains from 10 to 20 small black, shining, hard seeds.

[Illustration: Golden Seal Rootstock.]

Description of Rootstock--The fresh rootstock of Golden Seal, which has a rank, nauseating odor, is bright yellow, both internally and externally, with fibrous yellow rootlets produced from the sides. It is from 1 1/2 to 2 1/2 inches in length, from 1/4 to 3/4 of an inch in thickness, and contains a large amount of yellow juice.

In the dried state the rootstock is crooked, knotty and wrinkled, from 1 to 2 inches in length, and from one-eighth to one-third of an inch in diameter. It is a dull brown color on the outside and breaks with a clean, short, resinous fracture, showing a lemon-yellow color inside. After the rootstock has been kept for some time it will become greenish yellow or brown internally and its quality impaired. The cup-like depressions or stem scars on the upper surface of the rootstock resemble the imprint of a seal, whence the most popular name of the plant, golden seal, is derived. The rootstock as found in commerce is almost bare, the fibrous rootlets, which in drying become very wiry and brittle, breaking off readily and leaving only small protuberances.

The odor of the dried rootstock, while not so pronounced as in the fresh material, is peculiar, narcotic and disagreeable. The taste is exceedingly bitter, and when the rootstock is chewed there is a persistent acridity, which causes an abundant flow of saliva.

CHAPTER XX.

Collection, Prices and Uses--The root should be collected in autumn after the seeds have ripened, freed from soil, and carefully dried. After a dry season Golden Seal dies down soon after the fruit is mature, so that it often happens that by the end of September not a trace of the plant remains above ground; but if the season has been moist, the plant sometimes persists to the beginning of winter. The price of Golden Seal ranges from $1 to $1.50 a pound.

Golden Seal, which is official in the United States Pharmacopoeia, is a useful drug in digestive disorders and in certain catarrhal affections of the mucous membranes, in the latter instance being administered both internally and locally.

Cultivation--Once so abundant in certain parts of the country, especially in the Ohio Valley, Golden Seal is now becoming scarce thruout its range, and in consequence of the increased demand for the root, both at home and abroad, its cultivation must sooner or later be more generally undertaken in order to satisfy the needs of medicine. In some parts of the country the cultivation of Golden Seal is already under way.

The first thing to be considered in growing this plant is to furnish it, as nearly as possible, the conditions to which it has been accustomed in its native forest home. This calls for a well-drained soil, rich in humus, and partially shaded. Golden Seal stands transplanting well, and the easiest way to propagate it is to bring the plants in from the forest and transplant them to a properly prepared location, or to collect the rootstocks and to cut them into as many pieces as there are buds, planting these pieces in a deep, loose, well-prepared soil, and mulching, adding new mulch each year to renew the humus. With such a soil the cultivation of Golden Seal is simple and it will be necessary chiefly to keep down the weeds.

The plants may be grown in rows 1 foot apart and 6 inches apart in the row, or they may be grown in beds 4 to 8 feet wide, with walks between. Artificial shade will be necessary and this is supplied by the erection of lath sheds. The time required to obtain a marketable crop is from two to three years.

CHAPTER XXI.

COHOSH--BLACK AND BLUE.

Black Cohosh.

Cimicifuga Racemosa (L.) Nutt.

Synonym--Actaea Racemosa L.

Pharmacopoeial Name--Cimicifuga.

Other Common Names--Black snakeroot, bugbane, bugwort, rattlesnakeroot, rattleroot, rattleweed, rattletop, richweed, squawroot.

Habitat and Range--Altho preferring the shade of rich woods, black cohosh will grow occasionally in sunny situations in fence corners and woodland pastures. It is most abundant in the Ohio Valley, but it occurs from Maine to Wisconsin, south along the Allegheny Mountains to Georgia and westward to Missouri.

Description of Plant--Rising to a height of 3 to 8 feet, the showy, delicate-flowered spikes of the Black Cohosh tower above most of the other woodland flowers, making it a conspicuous plant in the woods and one that can be easily recognized.

[Illustration: Black Cohosh (Cimicifuga Racemosa) Leaves, Flowering Spikes and Rootstock.]

Black Cohosh is an indigenous perennial plant belonging to the same family as the Golden Seal, namely, the crowfoot family (Ranunculaceae). The tall stem, sometimes 8 feet in height, is rather slender and leafy, the leaves consisting of three leaflets, which are again divided into threes. The leaflets are about 2 inches long, ovate, sharp pointed at the apex, thin and smooth, variously lobed and the margins sharply toothed. The graceful, spikelike terminal cluster of flowers, which is produced from June to August, is from 6 inches to 2 feet in length. Attractive as these flower clusters are to the eye, they generally do not prove attractive very long to those who may gather them for their beauty, since the flowers emit an offensive odor, which account for some of the common names applied to this plant, namely, bugbane and bugwort, it having been thought that this odor was efficacious in driving away bugs. The flowers do not all open at one time and thus there may be seen buds, blossoms, and seed pods on one spike. The buds are white and globular and as they expand in flower there is practically nothing to the flower but very numerous white stamens and the pistil, but the stamens spread out around the pistil in such a manner as to give to the spike a somewhat feathery or fluffy appearance which is very attractive. The seed pods are dry, thick and leathery, ribbed, and about one-fourth of an inch long, with a small beak at the end. The smooth brown seeds are enclosed within the pods in two rows. Any one going thru the woods in winter may find the seed pods, full of seeds, still clinging to the dry, dead stalk, and the rattling of the seeds in the pods as the wind passes over them has given rise to the common names rattle-snakeroot (not "rattlesnake"-root), rattleweed, rattletop and rattleroot.

Description of Rootstock--The rootstock is large, horizontal and knotty or rough and irregular in appearance. The upper surface of the rootstock is covered with numerous round scars and stumps, the remains of former leaf stems, and on the fresh rootstocks may be seen the young, pinkish white buds which are to furnish the next season's growth. From the lower part of the rootstock long, fleshy roots arc produced. The fresh rootstock is very dark reddish brown on the outside, white within, showing a large central pith from which radiate rays of a woody texture, and on breaking the larger roots also the woody rays will be seen in the form of a cross. On drying, the rootstock becomes hard and turns much darker, both internally and externally, but the peculiar cross formation of the woody rays in both rootstock and roots, being lighter in color, is plainly seen without the aid of a magnifying glass. The roots in drying become wiry and brittle and break off very readily. Black cohosh has a heavy odor and a bitter, acrid taste.

Collection, Prices and Uses--The root should be collected after the fruit has ripened, usually in September. The price ranges from 2 to 3 cents a pound.

CHAPTER XXI. 153

The Indians had long regarded black cohosh as a valuable medicinal plant, not only for the treatment of snake bites, but it was also a very popular remedy among their women, and it is today considered of value as an alterative, emmenagogue, and sedative, and is recognized as official in the United States Pharmacopoeia.

Blue Cohosh.

Caulophyllum Thalictroides (L.) Michx.

Other Common Names--Caulophyllum, pappoose-root, squawroot, blueberryroot, blue ginseng, yellow ginseng.

[Illustration: Blue Cohosh (Caulophyllum Thalictroides).]

Habitat and Range--Blue Cohosh is found in the deep rich loam of shady woods from New Brunswick to South Carolina, westward to Nebraska, being abundant especially thruout the Allegheny Mountain region.

Description of Plant--This member of the barberry family (Berberidaceae) is a perennial herb, 1 to 3 feet in height, and indigenous to this country. It bears at the top one large, almost stemless leaf, which is triternately compound--that is, the main leaf stem divides into three stems, which again divide into threes, and each division bears three leaflets. Sometimes there is a smaller leaf, but similar to the other, at the base of the flowering branch. The leaflets are thin in texture, oval, oblong, or obovate and 3 to 5 lobed.

In the early stage of its growth this plant is covered with a sort of bluish green bloom, but it generally loses this and becomes smooth. The flowers are borne in a small terminal panicle or head, and are small and greenish yellow. They appear from April to May, while the leaf is still small. The globular seeds, which ripen about August, are borne on stout stalks in membranous capsules and resemble dark-blue berries.

Description of Rootstock--The thick, crooked rootstock of Blue Cohosh is almost concealed by the mass of matted roots which surrounds it. There are numerous cup-shaped scars and small branches on the upper surface of the rootstock, while the lower surface gives off numerous long, crooked, matted roots. Some of the scars are depressed below the surface of the rootstock, while others are raised above it. The outside is brownish and the inside tough and woody. Blue Cohosh possesses a slight odor and a sweetish, somewhat bitter and acrid taste. In the powdered state it causes sneezing.

Collection, Prices and Uses--The root is dug in the fall. Very often the roots of Golden Seal or twinleaf are found mixed with those of Blue Cohosh. The price of Blue Cohosh root ranges from 2 1/2 to 4 cents a pound.

Blue Cohosh, official in the United States Pharmacopoeia for 1890, is used as a demulcent, antispasmodic, emmenagogue and diuretic.

CHAPTER XXII.

SNAKEROOT--CANADA AND VIRGINIA.

Canada Snakeroot.

Asarum Canadense L.

Other Common Names--Asarum, wild ginger, Indian ginger, Vermont snakeroot, heart-snakeroot, southern snakeroot, black snakeroot, colt's-foot, snakeroot, black snakeweed, broadleaved asarabacca, false colt's-foot, cat's foot, colicroot.

Habitat and Range--This inconspicuous little plant frequents rich woods or rich soil along road sides from Canada south to North Carolina and Kansas.

Description of Plant--Canada snakeroot is a small, apparently stemless perennial, not more than 6 to 12 inches in height, and belongs to the birthwort family (Aristolochaceae). It usually has but two leaves which are borne on slender, finely hairy stems; they are kidney shaped or heart shaped, thin, dark green above and paler green on the lower surface, strongly veined, and from 4 to 7 inches broad.

The solitary bell-shaped flower is of an unassuming dull brown or brownish purple and this modest color, together with its position on the plant, renders it so inconspicuous as to escape the notice of the casual observer. It droops from a short, slender stalk produced between the two leaf stems and is almost hidden under the two leaves, growing so close to the ground that it is sometimes buried beneath old leaves, and sometimes the soil must be removed before the flower can be seen. It is bell shaped, wooly, the inside darker in color than the outside and of a satiny texture. The fruit which follows is in the form of a leathery 6-celled capsule.

[Illustration: Canada Snakeroot (Asarum Canadense).]

Description of Rootstock--Canada snakeroot has a creeping, yellowish rootstock, slightly jointed, with this rootlets produced from joints which

CHAPTER XXII. 156

occur about every half inch or so. In the drug trade the rootstock is usually found in pieces a few inches in length and about one-eighth of an inch in diameter. These are four-angled, crooked, brownish and wrinkled on the outside, whitish inside and showing a large central pith, hard and brittle and breaking with a short fracture. The odor is fragrant and the taste spicy and aromatic, and has been said to be intermediate between ginger and serpentaria.

Collection, Prices and Uses--The aromatic root of Canada snakeroot is collected in autumn and the price ranges from 10 to 15 cents a pound. It was reported as very scarce in the latter part of the summer of 1906. Canada Snakeroot, which was official in the United States Pharmacopoeia from 1820 to 1880, is used as an aromatic, diaphoretic and carminative.

Serpentaria.

(1) Aristolochia serpentaris L. and (2) Aristolochia reticulata Nutt.

Pharmacopoeial Name--Serpentaria.

[Illustration: Verginia Serpentaria (Aristolochia serpentaris).]

Other Common Names--(1) Virginia serpentaria, Virginia snakeroot, serpentary, snakeweed, pelican-flower, snagrel, sangrel, sangree-root; (2) Texas serpentaria, Texas snakeroot, Red River snakeroot.

Habitat and Range--Virginia serpentaria is found in rich woods from Connecticut to Michigan and southward, principally along the Alleghenies, and Texas serpentaria occurs in the Southwestern States, growing along river banks from Arkansas to Louisiana.

Description of Virginia Serpentaria--About midsummer the queerly shaped flowers of this native perennial are produced. They are very similar to those of the better known "Dutchman's-pipe," another species of this genus, which is quite extensively grown as an ornamental vine for covering porches and trellises. Virginia serpentaria and Texas serpentaria both belong to the birth wort family (Aristolochiaceae). The Virginia serpentaria is nearly erect, the

CHAPTER XXII. 157

slender, wavy stem sparingly branched near the base, and usually growing to about a foot in height, sometimes, however, even reaching 3 feet. The leaves are thin, ovate, ovate lance shaped or oblong lance shaped, and usually heart shaped at the base; they are about 2 1/2 inches long and about 1 or 1 1/2 inches in width. The flowers are produced from near the base of the plant, similar to its near relative, the Canada snakeroot. They are solitary and terminal, borne on slender, scaly branches, dull brownish purple in color, and of a somewhat leathery texture; the calyx tube is curiously bent or contorted in the shape of the letter S. The fruit is a roundish 6-celled capsule, about half an inch in diameter and containing numerous seeds.

Description of Texas Serpentaria--This species has a very wavy stem, with oval, heart-shaped, clasping leaves, which are rather thick and strongly reticulated or marked with a network of veins; hence the specific name reticulata. The entire plant is hairy, with numerous long, coarse hairs. The small, densely hairy purplish flowers are also produced from the base of the plant.

Description of Rootstock--Serpentaria has a short rootstock with many thin, branching, fibrous roots. In the dried state it is thin and bent, the short remains of stems showing on the upper surface and the under surface having numerous thin roots about 4 inches in length, all of a dull yellowish brown color, internally white. It has a very agreeable aromatic odor, somewhat like camphor, and the taste is described as warm, bitterish and camphoraceous.

The Texas serpentaria has a larger rootstock, with fewer roots less interlaced than the Virginia serpentaria.

Collection, Prices and Uses--The roots of serpentaria are collected in autumn. Various other roots are sometimes mixed with serpentaria, but as they are mostly high-priced drugs, such as golden seal, pinkroot, senega and ginseng, their presence in a lot of serpentaria is probably accidental, due simply to proximity of growth of these plants. Abscess-root (Polemonium Reptans L.) is another root with which serpentaria is often adulterated. It is very similar to serpentaria, except that it is nearly white. The price of serpentaria ranges from 35 to 40 cents a pound.

Serpentaria is used for its stimulant, tonic, and diaphoretic properties. Both species are official in the United States Pharmacopoeia.

CHAPTER XXIII.

POKEWEED.

Phytolacca Decandra L. a.

Synonym--Phytolacca Americana (L). a.

Pharmacopoeial Name--Phytolacca.

Other Common Names--Poke, pigeon-berry, garget, scoke, pocan, coakum, Virginia poke, inkberry, red inkberry, American nightshade, cancer-jalap, redweed.

Habitat and Range--Pokeweed, a common, familiar, native weed, is found in rich, moist soil along fence rows, fields, and uncultivated land from the New England States to Minnesota south to Florida and Texas.

Description of Plant--In Europe, where pokeweed has become naturalized from his country, it is regarded as an ornamental garden plant, and, indeed, it is very showy and attractive with its reddish purple stems, rich green foliage, and clusters of white flowers and dark-purple berries.

The stout, smooth stems, arising from a very large perennial root, attain a height of from 3 to 9 feet and are erect and branched, green at first, then reddish. If a piece of the stem is examined, the pith will be seen to be divided into disk-shaped parts with hollow spaces between them. The smooth leaves are borne on short stems and are about 5 inches long and 2 to 3 inches wide, ovate or ovate oblong, acute at the apex, and the margins entire. The long-stalked clusters of whitish flowers, which appear from July to September are from 3 to 4 inches in length, the flowers numerous and borne on reddish stems. In about two months the berries will have matured and assumed a rich dark-purple color. These smooth and shining purple berries are globular, flattened at both ends, and contain black seeds embedded in a rich crimson juice. This plant belongs to the pokeweed family (Phytolaccaceae).

CHAPTER XXIII.

a. Phytolacca Americana L. by right of priority should be accepted but P. Decandra L. is used in conformity with the Pharmacopoeia.

[Illustration: Pokeweed (Phytolacca Decandra), Flowering and Fruiting Branch.]

Description of Root--Pokeweed has a very thick, long, fleshy root, conical in shape and branches very much resembling that of horseradish and poisonous. In commerce it usually occurs in transverse or lengthwise slices, the outside a yellowish brown and finely wrinkled lengthwise and thickly encircled with lighter colored ridges. It breaks with a fibrous fracture and is yellowish gray within. The transverse slices show many concentric rings. There is a slight odor and the taste is sweetish and acrid. The root when powdered causes sneezing.

[Illustration: Pokeweek Root.]

Collection, Prices and Uses--The root of the Pokeweed, which is official in the United States Pharmacopoeia, is collected in the latter part of autumn, thoroughly cleaned, cut into a transverse or lengthwise slices, and carefully dried. It brings from 2 1/2 to 4 cents a pound.

The root is used for its alterative properties in treating various diseases of the skin and blood, and in certain cases in relieving pain and allaying inflammation. It also acts upon the bowels and causes vomiting.

The berries when fully matured are also used in medicine.

The young and tender shoots of the pokeweed are eaten in spring, like asparagus, but bad results may follow if they are not thoroughly cooked or if they are cut too close to the root.

CHAPTER XXIV.

MAY-APPLE.

Podophyllum Peltatum L.

Pharmacopoeial Name--Podophyllum.

Other Common Names--Mandrake, wild mandrake, American mandrake, wild lemon, ground-lemon, hog-apple, devil's-apple, Indian apple, raccoon-berry, duck's-foot, umbrella-plant, vegetable calomel.

Habitat and Range--The May-apple is an indigenous plant, found in low woods, usually growing in patches, from western Quebec to Minnesota, south to Florida and Texas.

Description of Plant--A patch of May-apple can be distinguished from afar, the smooth, dark-green foliage and close and even stand making it a conspicuous feature of the woodland vegetation.

May-apple is a perennial plant, and belongs to the barberry family (Berberidaceae.) It is erect and grows about 1 foot in height. The leaves are only two in number, circular in outline, but with five to seven deep lobes, the lobes 2 cleft, and toothed at the apex; they are dark green above, the lower surface lighter green and somewhat hairy or smooth, sometimes 1 foot in diameter, and borne on long leafstalks which are fixed to the center of the leaf, giving it an umbrella-like appearance. The waxy-white, solitary flower, sometimes 2 inches in diameter, appears in May, nodding on its short stout stalk, generally right between the two large umbrella-like leaves, which shade and hide it from view. The fruit which follows is lemon shaped, at first green, then yellow, about 2 inches in length and edible, altho when eaten immoderately it is known to have produced bad effects.

In a patch of May-apple plants there are always a number of sterile or flowerless stalks, which bear leaves similar to those of the flowering plants.

CHAPTER XXIV.

[Illustration: May-apple (Podophyllum Pellatum), Upper Portion of Plant with Flower and Rootstock.]

Description of Rootstock--The horizontally creeping rootstock of May-apple when taken from the ground, is from 1 to 6 feet or more in length, flexible, smooth, and round, dark brown on the outside and whitish and fleshy within; at intervals of a few inches are thickened joints, on the upper surface of which are round stem scars and on the lower side a tuft of rather stout roots. Sometimes the rootstock bears lateral branches. The dried rootstock, as it occurs in the stores, is in irregular, somewhat cylindrical pieces, smooth or somewhat wrinkled, yellowish brown or dark brown externally, whitish to pale brown internally, breaking with a short, sharp fracture, the surface of which is mealy. The odor is slight and the taste at first sweetish, becoming very bitter and acrid.

Collection, Prices and Uses--The proper time for collecting the rootstock is in the latter half of September or in October. The price paid for May-apple root ranges from 3 to 6 cents a pound.

May-apple root, which is recognized as official in the United States Pharmacopoeia, is an active cathartic and was known as such to the Indians.

CHAPTER XXV.

SENECA SNAKEROOT.

Polygala Senega L.

Pharmacopoeial Name--Senega.

Other Common Names--Senega snakeroot, Seneca-root, rattlesnake-root, mountain flax.

Habitat and Range--Rocky woods and hillsides are the favorite haunts of this indigenous plant. It is found in such situations from New Brunswick and western New England to Minnesota and the Canadian Rocky Mountains, and south along the Allegheny Mountains to North Carolina and Missouri.

Description of Plant--The perennial root of this useful little plant sends up a number of smooth, slender, erect stems (as many as 15 to 20 or more), sometimes slightly tinged with red, from 6 inches to a foot in height, and generally unbranched. The leaves alternate on the stem, are lance shaped or oblong lance shaped, thin in texture, 1 to 2 inches long, and stemless. The flowering spikes are borne on the ends of the stems and consist of rather crowded, small, greenish white, insignificant flowers. The flowering period of Seneca Snakeroot is from May to June. The spike blossoms gradually, and when the lower-most flowers have already fruited the upper part of the spike is still in flower. The seed capsules are small and contain two black, somewhat hairy seeds. The short slender stalks supporting these seed capsules have a tendency to break off from the main axis before the seed is fully mature, leaving the spike in a rather ragged-looking condition, and the yield of seed, therefore, is not very large. Seneca Snakeroot belongs to the milkwort family (Polygalaceae).

A form of Seneca Snakeroot, growing mostly in the North Central States and distinguished by its taller stems and broader leaves, has been called Polygala Senega Var. Latifolia.

CHAPTER XXV.

[Illustration: Seneca Snakeroot (Polygala Senega), Flowering Plant with Root.]

Description of Root--Seneca Snakeroot is described in the United States Pharmacopoeia as follows: "Somewhat cylindrical, tapering, more or less flexuous, 3 to 15 cm. long and 2 to 8 mm. thick, bearing several similar horizontal branches and a few rootlets; crown knotty with numerous buds and short stem remnants; externally yellowish gray or brownish yellow, longitudinally wrinkled, usually marked by a keel which is more prominent in perfectly dry roots near the crown; fracture short, wood light yellow, usually excentrically developed; odor slight, nauseating; taste sweetish, afterwards acrid."

The Seneca Snakeroots found in commerce vary greatly in size, that obtained from the South, which is really the official drug, being usually light colored and small. The principal supply of Seneca Snakeroot now comes from Minnesota, Wisconsin, and farther northward, and this western Seneca Snakeroot has a much larger, darker root, with a crown or head sometimes measuring 2 or 3 inches across and the upper part of the root very thick. It is also less twisted and not so distinctly keeled.

Seneca Snakeroot is often much adulterated with the roots of other species of Polygala and of other plants.

Collection, Prices and Uses--The time for collecting Seneca Snakeroot is in autumn. Labor conditions play a great part in the rise and fall of prices for this drug. It is said that very little Seneca Snakeroot has been dug in the Northwest during 1906, due to the fact that the Indians and others who usually engage in this work were so much in demand as farm hands and railroad laborers, which paid them far better than the digging of Seneca Snakeroot. Collectors receive from about 55 to 70 cents a pound for this root.

This drug, first brought into prominence as a cure for snake bite among the Indians, is now employed as an expectorant, emetic and diuretic. It is official in the Pharmacopoeia of the United States.

CHAPTER XXVI.

LADY'S-SLIPPER.

(1) Cypripedium hirsutum Mill and (2) Cypripedium parviflorum Salisb.

Synonym--(1) Cypripedium Pubescens Wild.

Pharmacopoeial Name--Cypripedium.

Other Common Names--(1) Large yellow lady's-slipper, yellow lady's-slipper, yellow moccasin-flower, Venus'-shoe, Venus'-cup, yellow Indian-shoe, American valerian, nerve-root, male nervine, yellow Noah's-ark, yellows, monkey-flower, umbil-root, yellow umbil; (2) small yellow lady's-slipper.

Habitat and Range--Both of these native species frequent bogs and wet places in deep shady woods and thickets. The large yellow lady's-slipper may be found from Nova Scotia south to Alabama and west to Nebraska and Missouri. The range for the small yellow lady's-slipper extends from Newfoundland south along the mountains to Georgia and west to Missouri, Washington and British Columbia.

Description of Plants--The orchid family (Orchicaceae), to which the lady's-slipper belong, boasts of many beautiful, showy and curious species and the lady's-slipper is no exception. There are several other plants to which the name lady's-slipper has been applied, but one glance at the peculiar structure of the flowers in the species under consideration, as shown in the illustration will enable any one to recognize them as soon as seen.

The particular species of lady's-slipper under consideration in this article do not differ very materially from each other. Both are perennials, growing from 1 to about 2 feet in height, with rather large leaves and with yellow flowers more or less marked with purple, the main difference being that in hirsutum the flower is larger and pale yellow, while in parviflorum the flower is small, bright yellow, and perhaps more prominently striped and spotted with purple. The stem, leaves and inside of corolla or lip are somewhat hairy in the large

CHAPTER XXVI.

yellow lady's-slipper, but not in the small yellow lady's-slipper. These hairs are said to be irritating to some people in whom they cause an eruption of the skin.

[Illustration: Large Yellow Lady's Slipper (Cyrpripedium Hirsutum).]

The leaves of the Lady's-Slipper vary in size from 2 to 6 inches in length and from 1 to 3 inches in width, and are broadly oval or elliptic, sharp pointed, with numerous parallel veins, and sheathing at the base, somewhat hairy in the large Lady's-Slipper. The solitary terminal flower, which appears from May to June, is very showy and curiously formed, the lip being the most prominent part. This lip looks like a large inflated bay (1 to 2 inches long in the large Lady's-Slipper), pale yellow or bright yellow in color, variously striped and blotched with purple. The other parts of the flower are greenish or yellowish, with purple stripes, and the petals are usually twisted.

Description of Rootstock--The Rootstock is of horizontal growth, crooked, fleshy and with numerous wavy, fibrous roots. As found in commerce, the rootstocks are from 1 to 4 inches in length, about an eighth of an inch in thickness, dark brown, the upper surface showing numerous round cup-shaped scars, the remains of former annual stems, and the lower surface thickly covered with wavy, wiry, and brittle roots, the latter breaking off with a short, white fracture. The odor is rather heavy and disagreeable and the taste is described as sweetish, bitter and somewhat pungent.

Collection, Prices and Uses--Both rootstock and roots are used and these should be collected in autumn, freed from dirt and carefully dried in the shade. These beautiful plants are becoming rare in many localities. Sometimes such high priced drugs as golden seal and senega are found mixed with the lady's-slipper, but as these are more expensive than the lady's-slipper it is not likely that they are included with fraudulent intent and they can be readily distinguished. The prices paid to collectors of this root range from 32 to 35 cents a pound.

The principal use of Lady's-Slipper, which is official in the United States Pharmacopoeia, is as an antispasmodic and nerve tonic, and it has been used for the same purposes as valerian.

CHAPTER XXVII.

FOREST ROOTS.

The facts set forth in the following pages are from American Root Drugs, a valuable pamphlet issued in 1907 by U. S. Department of Agriculture--Bureau of Plant Industry--and written by Alice Henkel.

Bethroot.

Trillium Erectum L.

Other Common Names: Trillium, red trillium, purple trillium, ill-scented trillium, birthroot, birthwort, bathwort, bathflower, red wake-robin, purple wake-robin, ill-scented wake-robin, red-benjamin, bumblebee-root, daffydown-dilly, dishcloth, Indian balm, Indian shamrock, nosebleed, squawflower, squawroot, wood-lily, true-love, orange-blossom. Many of these names are applied also to other species of Trillium.

Habitat and Range--Bethroot is a native plant growing in rich soil in damp, shady woods from Canada south to Tennessee and Missouri.

Description of Plant--This plant is a perennial belonging to the lily-of-the-valley family (Convallariaceae). It is a low growing plant, from about 8 to 16 inches in height, with a rather stout stem, having three leaves arranged in a whorl near the top. These leaves are broadly ovate, almost circular in outline, sharp pointed at the apex and narrowed at the base, 3 to 7 inches long and about as wide, and practically stemless.

Not only the leaves of this plant, but the flowers and parts of the flowers are arranged in threes, and this feature will serve to identify the plant. The solitary terminal flower of Bethroot has three sepals and three petals, both more or less lance shaped and spreading, the former greenish, and the petals, which are 1 1/4 inches long and one-half inch wide, are sometimes dark purple, pink, greenish, or white. The flower has an unpleasant odor. It appears from April to June and is followed later in the season by an oval, reddish berry.

CHAPTER XXVII.

[Illustration: Bethroot (Trillium Erectum).]

Various other species of Trillium are used in medicine, possessing properties similar to those of the species under consideration. These are also very similar in appearance to Trillium Erectum.

Description of Root--Bethroot, as found in the stores, is short and thick, of a light-brown color externally, whitish or yellowish inside, somewhat globular or oblong in shape, and covered all around with numerous pale brown, shriveled rootlets. The top of the root generally shows a succession of fine circles or rings, and usually bears the remains of stem bases.

The root has a slight odor, and is at first sweetish and astringent, followed by a bitter and acrid taste. When chewed it causes a flow of saliva.

Collection, Prices and Uses--Bethroot is generally collected toward the close of summer. The price ranges from 7 to 10 cents a pound.

It was much esteemed as a remedy among the Indians and early settlers. Its present use is that of an astringent, tonic, and alterative, and also that of an expectorant.

Culver's-Root.

Veronica Virginia L. (a)

Synonym--Leptandra Virginica (L) Nutt. (a)

Other Common Names--Culver's physic, blackroot, bowman's-root, Beaumont-root, Brinton-root, tall speedwell, tall veronica, physic-root, wholywort.

Habitat and Range--This common indigenous herb is found abundantly in moist, rich woods, mountain valleys, meadows and thickets from British Columbia south to Alabama, Missouri and Nebraska.

CHAPTER XXVII.

Description of Plant--Culver's-Root is a tall, slender stemmed perennial belonging to the figwort family (Scrophulariaceae). It is from 3 to 7 feet in height, with the leaves arranged around the simple stems in whorls of three to nine. The leaves are borne on very short stems, are lance shaped, long pointed at the apex, narrowed at the base, and sharply toothed, 3 to 6 inches in length and 1 inch or less in width. The white tube-shaped flowers, with two long protruding stamens, are produced from June to September and are borne in several terminal, densely crowded, slender, spikelike heads from 3 to 8 inches long.

(a) Some authors hold that this plant belongs to the genus Leptandra and that its name should be Leptandra virginica (L.) Nutt. The Pharmacopoeia is here followed.

[Illustration: Culver's Root (Veronica Virginica), Flowering Top and Rootstock.]

The flowers, as stated, are usually white, tho the color may vary from a pink to a bluish or purple and on account of its graceful spikes of pretty flowers it is often cultivated in gardens as an ornamental plant. The fruits are small, oblong, compressed, many-seeded capsules.

Description of Rootstock--After they are dried the rootstocks have a grayish brown appearance on the outside, and the inside is hard and yellowish, either with a hollow center or a brownish or purplish pith. When broken the fracture is tough and woody. The rootstock measures from 4 to 6 inches in length, is rather thick and bent, with branches resembling the main rootstock. The upper surface has a few stem scars, and from the sides and underneath numerous coarse, brittle roots are produced which have the appearance of having been artificially inserted into the rootstock. Culver's-root has a bitter and acrid taste, but no odor.

Collection, Price and Uses--The rootstock and roots should be collected in the fall of the second year. When fresh these have a faint odor resembling somewhat that of almonds, which is lost in the drying. The bitter, acrid taste of Culver's-root also becomes less the longer it is kept, and it is said that it should be kept at least a year before being used. The price paid to collectors ranges from 6 to 10 cents a pound.

Culver's-root, which is official in the United States Pharmacopoeia, is used as an alterative, cathartic and in disorders of the liver.

Stone-Root.

Collinsonia Canadensis L.

Other Common Names--Collinsonia, knob-root, knobgrass, knobweed, knotroot, horse-balm, horseweed, richweed, richleaf, ox-balm, citronella.

Habitat and Range--Stoneroot is found in moist, shady woods from Maine to Wisconsin, south to Florida and Kansas.

Description of Plant--Like most of the other members of the mint family (Menthaceae), Stoneroot is aromatic also, the fresh flowering plant possessing a very pleasant, lemon-like odor. It is a tall perennial herb, growing as high as 5 feet. The stem is stout, erect, branched, smooth, or the upper part hairy.

[Illustration: Stoneroot (Collinsonia Canadensis).]

The leaves are opposite, about 3 to 8 inches long, thin, ovate, pointed at the apex, narrowed or sometimes heart-shaped at the base, and coarsely toothed; the lower leaves are largest and are borne on slender stems, while the upper ones are smaller and almost stemless. Stoneroot is in flower from July to October, producing large, loose, open terminal panicles or heads of small, pale-yellow lemon-scented flowers. The flowers have a funnel-shaped 2-lipped corolla, the lower lip, larger, pendant and fringed, with two very much protruding stamens.

Description of Root--Even the fresh root of this plant is very hard. It is horizontal, large, thick, and woody, and the upper side is rough and knotty and branched irregularly. The odor of the root is rather disagreeable, and the taste pungent and spicy. In the fresh state, as well as when dry, the root is extremely hard, whence the common name "stoneroot." The dried root is grayish brown externally, irregularly knotty on the upper surface from the remains of branches and the

CHAPTER XXVII. 171

scars left by former stems and the lower surface showing a few thin roots. The inside of the root is hard and whitish.

Collection, Prices and Uses--Stoneroot, which is collected in autumn, is employed for its tonic, astringent, diuretic and diaphoretic effects. The price of the root ranges from 2 to 3 1/2 cents a pound.

The leaves are used by country people as an application to bruises.

Crawley-Root.

Corallorhiza Odontorhiza (Wild) Nutt.

Other Common Names--Corallorhiza, crawley, coralroot, small coralroot, small-flowered coralroot, late coralroot, dragon's-claw, chickentoe, turkey-claw, feverroot.

Habitat and Range--Rich, shady woods having an abundance of leaf mold produce this curious little plant. It may be found in such situations from Maine to Florida, westward to Michigan and Missouri.

Description of Plant--This peculiar native perennial, belonging to the orchid family (Orchidaceae) is unlike most other plants, being leafless, and instead of a green stem it has a purplish brown, sheathed scape, somewhat swollen or bulbous at the base and bearing a clustered head of purplish flowers 2 to 4 inches long. It does not grow much taller than about a foot in height.

The flowers, 6 to 20 in a head, appear from July to September, and consist of lance-shaped sepals and petals, striped with purple and a broad, whitish, oval lip, generally marked with purple and narrowed at the base. The seed capsule is large oblong, or some what globular.

[Illustration: Crawley-root (Corallorhiza Odontorhiza).]

Description of Rootstock--The rootstock of this plant is also curious, resembling in its formation a piece of coral on account of which it is known by the name of "coralroot." The other common names, such as chickentoe, turkey-claw, etc., all have reference to the form of the rootstock. As found in commerce, Crawley-root consists of small, dark-brown wrinkled pieces, the larger ones branched like coral. The taste at first is sweetish, becoming afterwards slightly bitter. It has a peculiar odor when fresh, but when dry it is without odor.

Collection, Prices and Uses--Crawley-root should be collected in July or August The price ranges from 20 to 50 cents a pound. Other species of Corallorhiza are sometimes collected and are said to probably possess similar properties. This root is aid to be very effective for promoting perspiration and it is also used as a sedative and in fever.

CHAPTER XXVIII.

FOREST PLANTS.

Male Fern.

Pharmacopoeial Name--Aspidium. Other Common Names: (1) Male shield-fern, sweet brake, knotty brake, basket-fern, bear's-paw root; (2) marginal-fruited shield-fern, evergreen wood-fern.

Habitat and Range--These ferns are found in rocky woods, the male shield-fern inhabiting the region from Canada westward to the Rocky Mountains and Arizona. It is widely distributed also through Europe, northern Asia, northern Africa, and South America. The marginal-fruited shield-fern, one of our most common ferns, occurs from Canada southward to Alabama and Arkansas.

Description of Plants--Both of these species are tall, handsome ferns, the long, erect fronds, or leaves, arising from a chaffy, scaly base, and consisting of numerous crowded stemless leaflets, which are variously divided and notched. There is but little difference between these two species. The male shield-fern is perhaps a trifle stouter, the leaves growing about 3 feet in length and having a bright-green color, whereas the marginal-fruited shield-fern has lighter green leaves, about 2 1/2 feet in length, and is of more slender appearance. The principal difference, however, is found in the arrangement of the "sori," or "fruit dots," These are the very small, round, tawny dots that are found on the backs of fern leaves, and in the male shield-fern these will be found arranged in short rows near the midrib, while in the marginal-fruited shield-fern, as this name indicates, the fruit dots are placed on the margins of the fronds. Both plants are perennials and members of the fern family (Polypodiaceae).

[Illustration: Marginal-Fruited Shield-Fern (Dryopteris Marginalis).]

Description of the Rootstock--These ferns have stout ascending or erect chaffy rootstocks, or rhizomes as they are technically known. As taken from the ground the rootstock is from 6 to 12 inches in length and 1 to 2 inches thick, covered with closely overlapping, brown, slightly curved stipe bases or leaf bases and soft, brown, chaffy scales. The inside of the rootstock is pale green. As found in the stores, however, male-fern with the stipe bases and roots removed measure about 3 to 6 inches in length and about one-half to 1 inch in thickness, rough where the stipe bases have been removed, brown outside, pale green and rather spongy inside.

The stipe bases remain green for a very long period, and these small, claw-shaped furrowed portions, or "fingers" as they are called, form a large proportion of the drug found on the American market and, in fact, are said to have largely superseded the rootstock. Male-fern has a disagreeable odor, and the taste is described as bitter-sweet, astringent, acrid, and nauseous.

Collection, Prices and Uses--The best time for collecting Male-fern root is from July to September. The root should be carefully cleaned, but not washed, dried out of doors in the shade as quickly as possible, and shipped to druggists at once. The United States Pharmacopoeia directs that "the chaff, together with the dead portions of the rhizome and stipes, should be removed, and only such portions used as have retained their internal green color."

Great care is necessary in the preservation of this drug in order to prevent it from deteriorating. If kept too long its activity will be impaired, and it is said that it will retain its qualities much longer if it is not peeled until required for use. The unreliability sometimes attributed to this drug can in most instances be traced to the presence of the rootstocks of other ferns with which it is often adulterated, or it will be found to be due to improper storing or to the length of time that it has been kept.

The prices paid for Male-fern root range from 5 to 10 cents a pound.

Male-fern, official in the United States Pharmacopoeia, has been used since the remotest times as a remedy for worms.

Grave results are sometimes caused by overdoses.

Goldthread.

CHAPTER XXVIII.

Coptis Trifolia (L.) Salisb.

Other Common Names--Coptis, cankerroot, mouthroot, yellowroot.

Habitat and Range--This pretty little perennial is native in damp, mossy woods and bogs from Canada and Alaska south of Maryland and Minnesota. It is most common in the New England States, northern New York and Michigan, and in Canada, where it frequents the dark sphagnum swamps, cold bogs and in the shade of dense forests of cedars, pines and other evergreens.

[Illustration: Goldthread (Coptis Trifolia).]

Description of Plant--Any one familiar with this attractive little plant will agree that it is well named. The roots of Goldthread, running not far beneath the surface of the ground, are indeed like so many tangled threads of gold. The plant in the general appearance of its leaves and flowers very closely resembles the strawberry plant. It is of low growth, only 3 to 6 inches in height, and belongs to the crowfoot family (Ranunculaceae). The leaves are all basal, and are borne on long, slender stems; they are evergreen, dark green and shining on the upper surface and lighter green beneath, divided into three parts, which are prominently veined and toothed. A single small, white, star-shaped flower is borne at the ends of the flowering stalks, appearing from May to August. The 5 to 7 sepals or lobes of the calyx are white and like petals, and the petals of the corolla, 5 to 7 in number, are smaller, club shaped, and yellow at the base. The seed pods are stalked, oblong, compressed, spreading, tipped with persistent style and containing small black seeds.

Description of Root--Goldthread has a long, slender, creeping root, which is much branched and frequently matted. The color of these roots is a bright golden yellow. As found in the stores, Goldthread consists usually of tangled masses of these golden-yellow roots, mixed with the leaves and stems of the plant, but the root is the part prescribed for use. The root is bitter and has no odor.

Collection, Prices and Uses--The time for collecting Goldthread is in autumn. After removing the covering of dead leaves and moss, the

CHAPTER XXVIII.

creeping yellow roots of Goldthread will be seen very close to the surface of the ground, from which they can be easily pulled. They should, of course, be carefully dried. As already stated, altho the roots and rootlets are the parts to be used, the commercial article is freely mixed with the leaves and stems of the plant. Evidences of the pine-woods home of this plant, in the form of pine needles and bits of moss, are often seen in the Goldthread received for market. Goldthread brings from 60 to 70 cents a pound.

The Indians and early white settlers used this little root as a remedy for various forms of ulcerated and sore mouth, and it is still used as a wash or gargle for affections of this sort. It is also employed as a bitter tonic.

Goldthread was official in the United States Pharmacopoeia from 1820 to 1880.

Twinleaf.

Jeffersonia Diphylla (L.) Pers.

Other Common Names--Jeffersonia, rheumatism-root, helmetpod, ground-squirrel pea, yellowroot.

Habitat and Range--Twinleaf inhabits rich, shady woods from New York to Virginia and westward to Wisconsin.

Description of Plant--This native herbaceous perennial is only about 6 to 8 inches in height when in flower. At the fruiting stage it is frequently 18 inches in height. It is one of our early spring plants, and its white flower, resembling that of bloodroot, is produced as early as April.

[Illustration: Twinleaf (Jeffersonia diphylla), Plant and Seed Capsule.]

The long-stemmed, smooth leaves, produced in pairs and arising from the base of the plant, are rather oddly formed. They are about 3 to 6 inches long, 2 to 4 inches wide, heart shaped or kidney shaped, but parted lengthwise into two lobes or divisions, really giving the appearance of two leaves; hence the common name "Twinleaf." The

CHAPTER XXVIII.

flower with its eight oblong, spreading white petals measures about 1 inch across, and is borne at the summit of a slender stalk arising from the root. The many-seeded capsule is about 1 inch long, leathery, somewhat pear shaped, and opening half way around near the top, the upper part forming a sort of lid. Twinleaf belongs to the barberry family. (Berberidaceae.)

Description of Rootstock--Twinleaf has a horizontal rootstock, with many fibrous, much-matted roots, and is very similar to that of blue cohosh, but not so long. It is thick, knotty, yellowish brown externally, with a resinous bark, and internally yellowish. The inner portion is nearly tasteless, but the bark has a bitter and acrid taste.

Collection, Prices and Uses--The rootstock is collected in autumn and is used as a diuretic, alterative, antispasmodic and a stimulating diaphoretic. Large doses are said to be emetic and smaller doses tonic and expectorant. The price paid for Twinleaf root ranges from about 5 to 7 cents a pound.

Canada Moonseed.

Menispermum Canadense L.

Other Common Names--Menispermum, yellow parilla, Texas sarsaparilla, yellow sarsarparilla, vine-maple.

Habitat and Range--Canada Moonseed is usually found along streams in woods, climbing over bushes, its range extending from Canada to Georgia and Arkansas.

[Illustration: Canada Moonseed (Menispermum Canadense).]

Description of Plant--This native perennial woody climber reaches a length of from 6 to 12 feet, the round, rather slender stem bearing very broad, slender-stalked leaves. These leaves are from 4 to 8 inches wide, smooth and green on the upper surface and paler beneath, roundish in outline and entire, or sometimes lobed and resembling the leaves of some of our maples, whence the common name "vine-maple" is probably derived. The bases of the leaves are generally heart shaped and the apex pointed or blunt. In July the loose clusters of small, yellowish or greenish white flowers are produced, followed in September by bunches of black one-seeded fruit, covered with a "bloom" and very much resembling grapes. Canada Moonseed belongs to the moonseed family (Menispermaceae.)

Description of Rootstock--The rootstock and roots are employed in medicine. In the stores it will be found in long, straight pieces, sometimes 3 feet in length, only about one-fourth of an inch in thickness, yellowish brown or grayish brown, finely wrinkled lengthwise, and giving off fine, hairlike, branched, brownish roots from joints which occur every inch or so. The inside shows a distinct white pith of variable thickness and a yellowish white wood with broad, porous wood rays, the whole breaking with a tough, woody fracture. It has practically no odor, but a bitter taste.

Collection, Prices and Uses--Canada Moonseed is collected in autumn and brings from 4 to 8 cents a pound. It is used as a tonic, alterative, and diuretic and was official in the United States Pharmacopoeia for 1890.

Wild Turnip.

Synonym--Arum Triphyllum L.

Other Common Names--Arum, three-leaved arum, Indian turnip, jack-in-the-pulpit, wake robin, wild pepper, dragon-turnip, brown dragon, devil's-ear, marsh-turnip, swamp-turnip, meadow-turnip, pepper-turnip, starch-wort, bog-onion, priest's-pintle and lords-and-ladies.

Habitat and Range--Wild Turnip inhabits moist woods from Canada to Florida and westward to Kansas and Minnesota.

Description of Plant--Early in April the quaint green and brownish purple hooded flowers of the wild turnip may be seen in the shady depths of the woods.

[Illustration: Wild Turnip (Arisaema Triphyllum).]

CHAPTER XXVIII.

It is a perennial plant belonging to the arum family (Araceae), and reaches a height of from 10 inches to 3 feet. The leaves, of which there are only one or two, unfold with the flowers; they are borne on long, erect, sheathing stalks, and consist of three smooth, oval leaflets, the latter are 3 to 6 inches long, and from 1 1/2 to 3 1/2 inches wide, net veined, and with one vein running parallel with the margins. The "flower" is curiously formed, somewhat like the calla lily, consisting of what is known botanically as a spathe, within which is inclosed the spadix. The spathe is an oval, leaflike part, the lower portion of which, in the flower under consideration, is rolled together so as to form a tube, while the upper, pointed part is usually bent forward, thus forming a flap of hood over the tube shaped part which contains the spadix. In fact it is very similar to the familiar flower of the calla lily of the gardens, except that, instead of being white, the wild turnip is either all green or striped with very dark purple, sometimes seeming almost black, and in the calla lily the "flap" is turned back, whereas in the wild turnip it is bent forward over the tube. Inside of the spathe is the spadix, also green or purple, which is club shaped, rounded at the summit, and narrowly contracted at the base, where it is surrounded by either the male or female flowers or both, in the latter case (the most infrequent) the male flowers being placed below the female flowers. In autumn the fruit ripens in the form of a bunch of bright scarlet, shining berries. The entire plant is acrid, but the root more especially so.

Description of the Root--The underground portion of this plant is known botanically as a "corm," and is somewhat globular and shaped like a turnip. The lower part of the corm is flat and wrinkled, while the upper part is surrounded by coarse, wavy rootlets. The outside is brownish gray and the inside white and mealy. It has no odor, but an intensely acrid, burning taste, and to those who may have been induced in their school days to taste of this root wild turnip will be familiar chiefly on account of its never-to-be-forgotten acrid, indeed, caustic, properties. The dried article of commerce consists of round, white slices, with brown edges, only slightly shrunken, and breaking with a starchy fracture.

Collection, Prices and Uses--The partially dried corm is used in medicine. It is dug in summer, transversely sliced, and dried. When first dug it is intensely acrid, but drying and heat diminish the acridity. It loses its acridity rapidly with age. Wild Turnip brings from 7 to 10 cents a pound.

The corm of Wild turnip, which was official in the United States Pharmacopoeia from 1820 to 1870, is used as a stimulant, diaphoretic, expectorant, and irritant.

CHAPTER XXIX.

THICKET PLANTS.

Black Indian Hemp.

Apocynum Cannabinum L.

Pharmacopoeial Name--Apocynum.

Other Common Names--Canadian hemp, American hemp, amy-root, bowman's-root, bitterroot, Indian-physic, rheumatism-weed, milkweed, wild cotton, Choctaw-root.

The name "Indian hemp" is often applied to this plant, but it should never be used without the adjective "black." "Indian hemp" is a name that properly belongs to Canabis indica, a true hemp plant, from which the narcotic drug "hashish" is obtained.

Habitat and Range--Black Indian hemp is a native of this country and may be found in thickets and along the borders of old fields thruout the United States.

Description of Plant--This is a common herbaceous perennial about 2 to 4 feet high, with erect or ascending branches, and, like most of the plants belonging to the dogbane family (Apocynaceae), contains a milky juice. The short-stemmed opposite leaves are oblong, lance shaped oblong or ovate-oblong, about 2 to 6 inches long, usually sharp pointed, the upper surface smooth and the lower sometimes hairy. The plant is in flower from June to August and the small greenish white flowers are borne in dense heads, followed later by the slender pods, which are about 4 inches in length and pointed at the apex.

Other Species--Considerable confusion seems to exist in regard to which species yields the root which has proved of greatest value medicinally. The Pharmacopoeia directs that "the dried rhizome and roots of Apocynum cannabinum or of closely allied species of Apocynum" be used.

CHAPTER XXIX.

[Illustration: Black Indian Hemp (Apocynum Cannabinum), Flowering Portion, Pods, and Rootstock.]

In the older botanical works and medical herbals only two species of Apocynum were recognized, namely, A. cannabinum L. and A. androsaemifolium L., altho it was known that both of these were very variable. In the newer botanical manuals both of these species still hold good, but the different forms and variations are now recognized as distinct species, those formerly referred to cannabium being distinguished by the erect or nearly erect lobes of the corolla, and those of the androsaemifolium group being distinguished by the spreading or recurved lobes of the corolla.

Among the plants that were formerly collected as Apocynum or varietal forms of it, and which are now considered as distinct species, may be mentioned in the following:

Riverbank-dogbane (A. Album Greene), which frequents the banks of rivers and similar moist locations from Maine to Wisconsin, Virginia and Missouri. This plant is perfectly smooth and has white flowers and relatively smaller leaves than A. cannabinum.

Velvet dogbane (A. pubescens R. Br.), which is common from Virginia to Illinois, Iowa and Missouri. The entire plant has a soft, hairy or velvety appearance, which renders identification easy. According to the latest edition of the National Standard Dispensatory it is not unlikely that this is the plant that furnishes the drug that has been so favorably reported upon.

Apocynum androsaemifolium is also gathered by drug collectors for Apocynum cannabinum. Its root is likewise employed in medicine, but its action is not the same as that of cannabinum and it should therefore not be substituted for it. It closely resembles cannabinum.

Description of Rootstock--The following description of the drug as found in commerce is taken from the United States Pharmacopoeia: "Of varying length, 3 to 8 mm. thick, cylindrical or with a few angles produced by drying, lightly wrinkled, longitudinally and usually more or less fissured transversely; orange-brown, becoming gray-brown on

CHAPTER XXIX. 183

keeping; brittle; fracture sharply transverse, exhibiting a thin brown layer of cork, the remainder of the bark nearly as thick as the radius of the wood, white or sometimes pinkish, starchy, containing laticiferous ducts; the wood yellowish, having several rings, finely radiate and very coarsely porous; almost inodorous, the taste starchy, afterwards becoming bitter and somewhat acrid."

Collection, Prices and Uses--The root of black Indian hemp is collected in autumn and brings from 8 to 10 cents a pound.

It is official in the United States Pharmacopoeia and has emetic, cathartic, diaphoretic, expectorant and diuretic properties, and on account of the last-named action it is used in dropsical affections.

The tough, fibrous bark of the stalks of Black Indian Hemp was employed by the Indians as a substitute for hemp in making twine, fishing nets, etc.

Chamaelirium, or Helonias.

Chamaelirium Luteum (L.) A. Gray.

Synonym--Helonias Dioica Pursh.

Other Common Names--Unicorn root, false unicorn-root, blazing star, drooping starwort, starwort, devil's-bit, unicorn's-horn.

In order to avoid the existing confusion of common names of this plant, it is most desirable to use the scientific names Chamaelirium or Helonias exclusively. Chamaelirium is the most recent botanical designation and will be used thruout this article, but the synonym Helonias is a name very frequently employed by the drug trade. The plant with which it is so much confused, Aletris farinosa, will also be designated thruout by its generic name, Aletris.

[Illustration: Chamaelirium (Chamaelirium Luteum).]

Habitat and Range--This native plant is found in open woods from Massachusetts to Michigan, south to Florida and Arkansas.

CHAPTER XXIX.

Description of Plant--Chamaelirium and Aletris (Aletris farinosa) have long been confused by drug collectors and others, owing undoubtedly to the transposition of some of their similar common names, such as "starwort" and "stargrass." The plants can scarcely be said to resemble each other, however, except perhaps in their general habit of growth.

The male and female flowers of Chamaelirium are borne on separate plants, and in this respect are entirely different from Aletris; neither do the flowers resemble those of Aletris.

Chamaelirium is an erect, somewhat fleshy herb, perennial, and belongs to the bunchflower family (Melanthiaceae.) The male plant grows to a height of from 1 1/2 to 2 1/2 feet, and the female plant is sometimes 4 feet tall and is also more leafy.

The plants have both basal and stem leaves, where as Aletris has only the basal leaves. The basal leaves of Chamaelirium are broad and blunt at the top, narrowing toward the base into a long stem; they are sometimes so much broadened at the top that they may be characterized as spoon shaped, and are from 2 to 8 inches long and from one-half to 1 1/2 inches wide. The stem leaves are lance shaped and sharp pointed, on short stems or stemless.

The white starry flowers of Chamaelirium are produced from June to July, those of the male plant being borne in nodding, graceful, plume-like spikes 3 to 9 inches long, and those of the female plant in erect spikes. The many seeded capsule is oblong, opening by three valves at the apex.

Another species is now recognized, Chamaelirium obovale Small, which seems to differ chiefly in having larger flowers and obovoid capsules.

Description of Rootstock--The rootstock of Chamaelirium does not in the least resemble that of Aletris, with which it is so generally confused. It is from one-half to 2 inches in length, generally curved upward at one end in the form of a horn (whence the common name, "unicorn") and having the appearance of having been bitten off. It is of a dark brown color with fine transverse wrinkles, rough, on the upper surface showing a few stem scars, and giving off from all sides numerous brown fibrous rootlets. The more recent rootlets have a soft outer covering, which in the older rootlets has worn away, leaving the fine but tough and woody whitish center. The rootlets penetrate to the central part of the rootstock, and this serves as a distinguishing character from Aletris, as a transverse section of Chamaelirium very plainly shows these fibers extending some distance within the rootstock. Furthermore, the rootstock of Chamaelirium exhibits a number of small holes wherever these rootlets have broken off, giving it the appearance of having become "wormy." It is hard and horny within and has a peculiar odor and a very bitter, disagreeable taste, whereas Aletris is not at all bitter.

Collection, Prices and Uses--Chamaelirium should be collected in autumn. The prices paid to collectors may be said to range from about 30 to 45 cents a pound. In the fall of 1906 a scarcity of this root was reported. As already indicated, Chamaelirium and Aletris are often gathered and mistaken for each other by collectors, but, as will be seen from the preceding description, there is really no excuse for such error.

From the confusion that has existed properties peculiar to the one plant have also been attributed to the other, but it seems now generally agreed that Chamaelirium is of use especially in derangements of women.

Wild Yam.

Dioscorea Villosa L.

Other Common Names--Dioscorea, colicroot, rheumatism-root, devil's bones.

Habitat and Range--Wild yam grows in moist thickets, trailing over adjacent shrubs and bushes, its range extending from Rhode Island to Minnesota, south to Florida and Texas. It is most common in the central and southern portions of the United States.

CHAPTER XXIX.

Description of Plant--This native perennial vine is similar to and belongs to the same family as the well-known cinnamon vine of the gardens--namely, the yam family (Dioscoreaceae.) It attains a length of about 15 feet, the stem smooth, the leaves heart shaped and 2 to 6 inches long by 1 to 4 inches wide.

[Illustration: Wild Yam (Dioscorea Villosa).]

The leaves, which are borne on long, slender stems, are thin, green, and smooth on the upper surface, paler and rather thickly hairy on the under surface. The small greenish yellow flowers are produced from June to July, the male flowers borne in drooping clusters about 3 to 6 inches long, and the female flowers in drooping spikelike heads. The fruit, which is in the form of a dry, membranous, 3-winged, yellowish green capsule, ripens about September and remains on the vine for some time during the winter.

Growing farther south than the species above mentioned is a variety for which the name Glabra has been suggested.

According to C. G. Lloyd, there is a variety of Dioscorea Villosa, the root of which first made its appearance among the true yam roots of commerce, and which was so different in form that it was rejected as an adulteration. The plant, however, from which the false root was derived was found upon investigation to be almost identical with the true yam, except that the leaves were perfectly smooth, lacking the hairiness on the under surface of the leaf which is characteristic of the true wild yam. The false variety also differs in its habit of growth, not growing in dense clumps like the true wild yam, but generally isolated. The root of the variety, however, is quite distinct from that of the true wild yam, being much more knotty. Lloyd states further that the hairiness or lack of hairiness on the under side of the leaf is a certain indication as to the form of the root.

Lloyd, recognizing the necessity of classifying these two yam roots of commerce, has designated the smooth-leaved variety as Dioscorea Villosa var. Glabra.

CHAPTER XXIX.

Description of Rootstocks--The rootstock of the true wild yam runs horizontally underneath the surface of the ground. As found in commerce, it consists of very hard pieces, 6 inches and sometimes 2 feet in length, but only about one-fourth or one-half of an inch in diameter, twisted, covered with a thin, brown bark, whitish within and showing stem scars almost an inch apart on the upper surface, small protuberances on the sides, and numerous rather wiry rootlets on the lower surface.

The false wild yam, on the other hand, has a much heavier, rough, knotty rootstock, with thick branches from 1 inch to 3 inches long, the upper surface covered with crowded stem scars and the lower side furnished with stout, wiry rootlets. Within it is similar to the true yam root.

Collection, Prices and Uses--The roots are generally collected in autumn, and bring from 2 1/2 to 4 cents a pound. Wild Yam is said to possess expectorant properties and to promote perspiration, and in large doses providing emetic. It has been employed in bilious colic, and by the negroes in the South in the treatment of muscular rheumatism.

CHAPTER XXX.

SWAMP PLANTS.

Skunk-Cabbage.

Synonyms--Dracontium Foetidum L.

Other Common Names--Dracontium, skunk-weed, polecat-weed, swamp-cabbage, meadow-cabbage, collard, fetid, hellebone, stinking poke, pockweed.

Habitat and Range--Swamps and other wet places from Canada to Florida, Iowa and Minnesota abound with this ill-smelling herb.

Description of Plant--Most of the common names applied to this plant, as well as the scientific names, are indicative of the most striking characteristic of this early spring visitor, namely, the rank, offensive, carrion odor that emanates from it. Skunk-Cabbage is one of the very earliest of our spring flowers, appearing in February or March, but it is safe to say that it is not likely to suffer extermination at the hand of the enthusiastic gatherer of spring flowers. In the latitude of Washington Skunk-Cabbage has been known to be in flower in December.

It is a curious plant, with its hood shaped, purplish striped flowers appearing before the leaves. It belongs to the arum family (Araceae) and is a perennial. The "flower" is in the form of a thick, ovate, swollen spathe, about 3 to 6 inches in height, the top pointed and curved inward, spotted and striped with purple and yellowish green. The spathe is not like that of the wild turnip or calla lily, to which family this plant also belongs, but the edges are rolled inward, completely hiding the spadix. In this plant the spadix is not spike-like, as in the wild turnip, but is generally somewhat globular, entirely covered with numerous, dull-purple flowers. After the fruit has ripened the spadix will be found to have grown considerably, the spathe meantime having decayed.

The leaves, which appear after the flower, are numerous and very large, about 1 to 3 feet in length and about 1 foot in width; they are thin

CHAPTER XXX.

in texture, but prominently nerved with fleshy nerves, and are borne on deeply channeled stems.

[Illustration: Skunk Cabbage (Spathyema Foetida).]

Description of Rootstock--Skunk-Cabbage has a thick, straight, reddish brown rootstock, from 3 to 5 inches long, and about 2 inches in diameter, and having a whorl of crowded fleshy roots which penetrate the soil to considerable depth. The dried article of commerce consists of either the entire rootstock and roots, which are dark brown and wrinkled within, or of very much compressed, wrinkled, transverse slices.

When bruised, the root has the characteristic fetid odor of the plant and possesses a sharp acrid taste, both of which become less the longer the root is kept.

Collection, Prices and Uses--The rootstock of Skunk-Cabbage are collected early in spring, soon after the appearance of the flower, or after the seeds have ripened, in August or September. It should be carefully dried, either in its entire state or deprived of the roots and cut into transverse slices. Skunk-Cabbage loses its odor and acridity with age, and should therefore not be kept longer than one season. The range of prices is from 4 to 7 cents a pound.

Skunk-Cabbage, official from 1820 to 1880, is used in affections of the respiratory organs, in nervous disorders, rheumatism, and dropsical complaints.

American Hellebore.

Veratrum Viride Ait.

Pharmacopoeial Name--Veratrum.

Other Common Names--True veratrum, green veratrum, American veratrum, green hellebore, swamp-hellebore, big hellebore, false hellebore, bear-corn, bugbane, bugwort, devil's-bite, earth-gall, Indian poke, itchweed, tickleweed, duckretter.

CHAPTER XXX.

Habitat and Range--American Hellebore is native in rich, wet woods, swamps and wet meadows. Its range extending from Canada, Alaska, and Minnesota south to Georgia.

Description of Plant--Early in spring, usually in company with the Skunk-Cabbage, the large bright green leaves of American Hellebore make their way thru the soil, their straight, erect leaf spears forming a conspicuous feature of the yet scanty spring vegetation. Later in the season a stout and erect leafy stem is sent up, sometimes growing as tall as 6 feet. It is solid and round, pale green, very leafy, and closely surrounded by the sheathing bases of the leaves, unbranched except in the flowering head. The leaves are hairy, prominently nerved, folded or pleated like a fan. They have no stems, but their bases encircle or sheathe the main stalk, and are very large, especially the lower ones, which are from 6 to 12 inches in length, from 3 to 6 inches in width, and broadly oval. As they approach the top of the plant the leaves become narrower. The flowers, which appear from May to July, are greenish yellow and numerous, and are borne in rather open clusters. American Hellebore belongs to the bunchflower family (Melanthiaceae) and is a perennial.

This species is a very near relative of the European white hellebore (Veratrum album L.), and in fact has by some been regarded as identical with it, or at least as a variety of it. It is taller than V. album and has narrower leaves and greener flowers. Both species are official in the United States Pharmacopoeia.

[Illustration: American Hellebore (Veratrum Viride).]

Description of Rootstock--The fresh rootstock of American Hellebore is ovoid or obconical, upright, thick, and fleshy, the upper part of it arranged in layers, the lower part of it more solid, and producing numerous whitish roots from all sides. In the fresh state it has a rather strong, disagreeable odor. As found in commerce, American Hellebore rootstock is sometimes entire, but more generally sliced, and is of a light brown or dark brown color externally and internally yellowish white. The roots, which are from 4 to 8 inches long, have a shriveled appearance, and are brown or yellowish. There is no odor to the dried rootstock, but when powdered it causes violent sneezing. The

CHAPTER XXX.   191

rootstock, which has a bitter and very acrid taste, is poisonous.

Collection, Prices and Uses--American Hellebore should be dug in autumn after the leaves have died and washed and carefully dried, either in the whole state or sliced in various ways. It deteriorates with age, and should therefore not be kept longer than a year.

The adulterations sometimes met with are the rootstocks of related plants, and the skunk-cabbage is also occasionally found mixed with it, but this is probably unintentional, as the two plants usually grow close together.

Collectors of American Hellebore root receive from about 3 to 10 cents a pound.

American Hellebore, official in the United States Pharmacopoeia, is an acrid, narcotic poison, and has emetic, diaphoretic, and sedative properties.

Water-Eryngo.

Eryngium Yuccifolium Michx.

Synonym--Eryngium aquaticum. L.

Other Common Names--Eryngium, eryngo, button-snakeroot, corn-snakeroot, rattlesnake-master, rattlesnake-weed, rattlesnake-flag.

[Illustration: Water-Eryngo (Eryngium Yuccifolium).]

Habitat and Range--Altho sometimes occurring on dry land, Water-Eryngo usually inhabits swamps and low, wet ground, from the pine barrens of New Jersey westward to Minnesota and south to Texas and Florida.

Description of Plant--The leaves of this plant are grasslike in form, rigid, 1 to 2 feet long and about one-half inch or a trifle more in width; they are linear, with parallel veins, pointed, generally clasping at the base, and the margins briskly soft, slender spines. The stout, furrowed stem reaches a height of from 2 to 6 feet and is generally unbranched except near the top. The insignificant whitish flowers are borne in dense, ovate-globular, stout-stemmed heads, appearing from June to September, and the seed heads that follow are ovate and scaly. Water-Eryngo belongs to the parsley family (Apiaceae) and is native in this country.

Description of Rootstock--The stout rootstock is very knotty, with numerous short branches, and produces many thick, rather straight roots, both rootstock and roots of a dark brown color, the latter wrinkled lengthwise. The inside of the rootstock is yellowish white. Water-Eryngo has a somewhat peculiar, slightly aromatic odor, and a sweetish mucilaginous taste at first, followed by some bitterness and pungency.

Collection, Prices and Uses--The root of this plant is collected in autumn and brings from 5 to 10 cents a pound.

Water-Eryngo is an old remedy and one of its early uses, as the several common names indicate, was for the treatment of snake bites. It was official in the United States Pharmacopoeia from 1820 to 1860, and is employed now as a diuretic and expectorant and for promoting perspiration. In large doses it acts as an emetic and the root, when chewed, excites a flow of saliva. It is said to resemble Seneca snakeroot in action.

Yellow Jasmine or Jessamine.

Gelsemium Sempervirens (L.) Ait. f.

Pharmacopoeial Name--Gelsemium.

Other Common Names--Carolina jasmine or jessamine, Carolina wild woodbine, evening trumpet-flower.

Habitat and Range--Yellow jasmine is a plant native to the South, found along the banks of streams, in woods, lowlands, and thickets, generally near the coast, from the eastern part of Virginia to Florida and Texas, south to Mexico and Guatemala.

CHAPTER XXX.

Description of Plant--This highly ornamental climbing or trailing plant is abundantly met with in the woods of the Southern states, its slender stems festooned over trees and fences and making its presence known by the delightful perfume exhaled by its flowers, filling the air with fragrance that is almost overpowering wherever the yellow jasmine is very abundant.

[Illustration: Yellow Jasmine (Gelsensium Sempervirens).]

The smooth, shining stems of this beautiful vine sometimes reach a length of 20 feet. The leaves are evergreen, lance shaped, entire, 1 1/2 to 3 inches long, rather narrow, borne on short stems, and generally remaining on the vine during the winter. The flowers, which appear from January to April, are bright yellow, about 1 to 1 1/2 inches long, the corolla funnel shaped. They are very fragrant but poisonous, and it is stated the eating of honey derived from jasmine flowers has brought about fatal results.

Yellow Jasmine is a perennial and belongs to a family that is noted for its poisonous properties, namely, the Logania family (Loganiaceae), which numbers among its members such powerful poisonous agents as the strychnine-producing tree.

Description of Rootstock--The rootstock of the Yellow Jasmine is horizontal and runs near the surface of the ground, attaining great length, 15 feet or more; it is branched, and here and there produces fibrous rootlets. When freshly removed from the ground it is very yellow, with a peculiar odor and bitter taste. For the drug trade it is generally cut into pieces varying from 1 inch to 6 inches in length, and when dried consists of cylindrical sections about 1 inch in thickness, the roots, of course, thinner. The bark is thin, yellowish brown, with fine silky bast fibers and the wood is tough and pale yellow, breaking with a splintery fracture and showing numerous fine rays radiating from a small central pith. Yellow Jasmine has a bitter taste and a pronounced heavy odor.

Collection, Prices and Uses--The root of Yellow Jasmine is usually collected just after the plant has come into flower and is cut into pieces from 1 to 6 inches long. It is often adulterated with portions of the stems, but these can be distinguished by their thinness and dark purplish color. The prices range from 3 to 5 cents a pound.

Yellow Jasmine, which is official in the United States Pharmacopoeia, is used for its powerful effect on the nervous system.

Sweet-Flag.

Acorus Calamus L.

Pharmacopoeial Name--Calamus.

Other Common Names--Sweet cane, sweet grass, sweet myrtle, sweet rush, sweet sedge, sweet segg, sweetroot, cinnamon-sedge, myrtle-flag, myrtle-grass, myrtle-sedge, beewort.

Habitat and Range--This plant frequents wet and muddy places and borders on streams from Nova Scotia to Minnesota, southward to Florida and Texas, also occurring in Europe and Asia. It is usually partly immersed in water, and is generally found in company with the cat-tail and other water-loving species of flag.

[Illustration: Sweet Flag (Acorus Calamus).]

Description of Plant--The sword like leaves of the Sweet-Flag resemble those of other flags so much that before the plant is in flower it is difficult to recognize simply by the appearance of its leaves. The leaves of the blue flag or "poison-flag," as it has been called, are very similar to those of the Sweet-Flag, and this resemblance often leads to cases of poisoning among children who thus mistake one for the other. However, as the leaves of the Sweet-Flag are fragrant, the odor will be a means of recognizing it. Of course when the Sweet-Flag is in flower the identification of the plant is easy.

The sheathing leaves of this native perennial, which belongs to the arum family (Araceae), are from 2 to 6 feet in height and about 1 inch in width; they are sharp pointed and have a ridged midrib running their entire length. The flowering head, produced from the side of the stalk, consists of a fleshy spike sometimes 3 1/2 inches long and about one-half inch in thickness, closely covered with very small, greenish yellow flowers, which appear from May to July.

Description of Rootstock--The long, creeping rootstock of the Sweet-Flag is thick and fleshy, somewhat spongy, and producing numerous rootlets. The odor is aromatic and agreeable, and taste pungent and bitter. The dried article, as found in the stores, consists of entire or split pieces of various lengths from 3 to 6 inches, light brown on the outside with blackish spots, sharply wrinkled lengthwise, the upper surface marked obliquely with dark leaf scars, and the lower surface showing many small circular scars, which, at first glance, give one the impression that the root is worm-eaten, but which are the remains of rootlets that have been removed from the rootstock. Internally the rootstock is whitish and of a spongy texture. The aromatic odor and pungent, bitter taste are retained in the dried article.

Collection, Prices and Uses--The United States Pharmacopoeia directs that the unpeeled rhizome, or rootstock, be used. It is collected either in early spring or late in autumn. It is pulled or grubbed from the soft earth, freed from adhering dirt, and the rootlets removed, as these are not so aromatic and more bitter. The rootstock is then carefully dried, sometimes by means of moderate heat. Sweet-Flag deteriorates with age and is subject to the attacks of worms. It loses about three-fourths of its weight in drying.

Some of the Sweet-Flag found in commerce consists of handsome white pieces. These usually come from Germany, and have been peeled before drying, but they are not so strong and aromatic as the unpeeled roots. Unpeeled Sweet-Flag brings from 3 to 6 cents a pound.

Sweet-Flag is employed as an aromatic stimulant and tonic in feeble digestion. The dried root is frequently chewed for the relief of dyspepsia.

Blue Flag.

Iris Versicolor L.

CHAPTER XXX.

Other Common Names--Iris, flag-lily, liver-lily, snake-lily, poison-flag, water-flag, American fleur-de-lis or flower-deluce.

Habitat and Range--Blue Flag delights in wet, swampy localities, making its home in marshes, thickets, and wet meadows from Newfoundland to Manitoba, south to Florida and Arkansas.

Description of Plant--The flowers of all of the species belonging to this genus are similar, and are readily recognized by their rather peculiar form, the three outer segments or parts reflexed or turned back and the three inner segments standing erect.

Blue Flag is about 2 to 3 feet in height, with an erect stem sometimes branched near the top, and sword shaped leaves which are shorter than the stem, from one-half to 1 inch in width, showing a slight grayish "bloom" and sheathing at the base. This plant is a perennial belonging to the iris family (Iridaceae), and is a native of this country. June is generally regarded as the month for the flowering of the Blue Flag, altho it may be said to be in flower from May to July, depending on the locality. The flowers are large and very handsome, each stem bearing from two to six or more. They consist of six segments or parts, the three outer ones turned back and the three inner ones erect and much smaller. The flowers are usually purplish blue, the "claw" or narrow base of the segments, variegated with yellow, green, or white and marked with purple veins.

All of the species belonging to this genus are more or less variegated in color; hence the name "iris," meaning "rainbow," and the specific name "versicolor," meaning "various colors." The name poison-flag has been applied to it on account of the poisonous effect it has produced in children, who, owing to the close resemblance of the plants before reaching the flowering stage, sometimes mistake it for sweet flag.

The seed capsule is oblong, about 1 1/2 inches and contains numerous seeds.

[Illustration: Blue Flag (Iris Versicolor).]

Description of Rootstock--Blue Flag has a thick, fleshy, horizontal rootstock, branched, and producing long, fibrous roots. It resembles sweet-flag (Calamus) and has been mistaken for it. The sections of the rootstock of Blue Flag, however, are flattened above and rounded below; the scars of the leaf sheaths are in the form of rings, whereas in sweet-flag the rootstock is cylindrical and the scars left by the leaf sheaths are obliquely transverse. Furthermore, there is a difference in the arrangement of the roots on the rootstock, the scars left by the roots in Blue Flag being close together generally nearer the larger end, while in sweet-flag the disposition of the roots along the rootstock is quite regular. Blue Flag is grayish brown on the outside when dried, and sweet-flag is light brown or fawn colored. Blue Flag has no well-marked odor and the taste is acrid and nauseous, and in sweet-flag there is a pleasant odor and bitter, pungent taste.

Collection, Prices and Uses--Blue Flag is collected in autumn and usually brings from about 7 to 10 cents a pound. Great scarcity of Blue Flag root was reported from the producing districts in the autumn of 1906. It is an old remedy, the Indians esteeming it highly for stomach troubles, and it is said that it was sometimes cultivated by them in near-by ponds on account of its medicinal value. It has also been used as a domestic remedy and is regarded as an alterative, diuretic and purgative. It was official in the United States Pharmacopoeia of 1890.

Crane's-Bill.

Geranium Maculatum L.

Pharmacopoeial Name--Geranium.

Other Common Names--Spotted crane's-bill, wild crane's-bill, stork's-bill, spotted geranium, wild geranium, alum-root, alumbloom, chocolate-flower, crowfoot, dovefoot, old-maid's-nightcap, shameface.

Habitat and Range--Crane's-Bill flourishes in low grounds and open woods from Newfoundland to Manitoba, south to Georgia and Missouri.

CHAPTER XXX.

Description of Plant--This pretty perennial plant belongs to the geranium family (Geraniaceae) and will grow sometimes to a height of 2 feet, but more generally it is only about a foot in height. The entire plant is more or less covered with hairs, and is erect and usually unbranched. The leaves are nearly circular or somewhat heart shaped in outline, 3 to 6 inches wide, deeply parted into three or five parts, each division again cleft and toothed. The basal leaves are borne on long stems, while those above have short stems. The flowers, which appear from April to June, are borne in a loose cluster; they are rose purple, pale or violet in color, about 1 inch or 1 1/2 inches wide, the petals delicately veined and woolly at the base and the sepals or calyx lobes with a bristle-shaped point, soft-hairy, the margins having a fringe of more bristly hairs. The fruit consists of a beaked capsule, springing open elastically, and dividing into five cells, each cell containing one seed.

[Illustration: Crane's-bill (Geranium Maculatum), Flowering Plant, Showing also Seed Pods and Rootstock.]

Description of Rootstock--When removed from the earth the rootstock of Crane's-bill is about 2 to 4 inches long, thick, with numerous branches bearing the young buds for next season's growth and scars showing the remains of stems of previous years, brown outside, white and fleshy internally, and with several stout roots. When dry, the rootstock turns a darker brown, is finely wrinkled externally, and has a rough spiny appearance, caused by the shrinking of the buds and branches and the numerous stem scars with which the root is studded. Internally it is of a somewhat purplish color. Crane's-bill root is without odor and the taste is very astringent.

Collection, Prices and Uses--Crane's-bill root depends for its medicinal value on its astringent properties and as its astringency is due to the tannin content, the root should, of course, be collected at that season of the year when it is richest in that constituent. Experiments have proved that the yield of tannin in Crane's-bill is greatest just before flowering, which is in April or May, according to locality. It should, therefore, be collected just before the flowering periods, and not, as is commonly the case, in autumn. The price of this root ranges from 4 to 8 cents a pound.

CHAPTER XXX.

Crane's-bill root, which is official in the United States Pharmacopoeia, is used as a tonic and astringent.

CHAPTER XXXI.

FIELD PLANTS.

Dandelion.

Taraxacum Officinale Weber, (a).

Synonyms--Taraxacum taraxacum (L.) Karst: (a) Taraxacum densleonis Desf.

Pharmacopoeial Names--Taraxacum.

Other Common Names--Blow-ball, cankerwort, doon-head, clock, fortune-teller, horse gowan, Irish daisy, yellow gowan, one-o'clock.

Habitat and Range--With the exception, possibly, of a few localities in the South, the dandelion is at home almost everywhere in the United States, being a familiar weed in meadows and waste places, and especially in lawns. It has been naturalized in this country from Europe and is distributed as a weed in all civilized parts of the world.

Description of Plant--It is hardly necessary to give a description of the dandelion, as almost every one is familiar with the coarsely toothed, smooth, shining green leaves, the golden-yellow flowers which open in the morning and only in fair weather, and the round fluffy seed heads of this only too plentiful weed of the lawns. In spring the young, tender leaves are much sought after by the colored market women about Washington, who collect them by the basketful and sell them for greens and salad.

Dandelion is a perennial belonging to the chicory family (Cichoriaceae) and is in flower practically throughout the year. The entire plant contains a white milky juice.

Description of Root--The dandelion has a large, thick and fleshy taproot, sometimes measuring 20 inches in length. In commerce, dandelion root is usually found in pieces 3 to 6 inches long, dark brown on the outside and strongly wrinkled lengthwise. It breaks with a short

CHAPTER XXXI.

fracture and shows the thick whitish bark marked with circles of milk ducts and a thin woody center, which is yellow and porous. It is practically without odor and has a bitter taste.

[Illustration: Dandelion (Taraxacum Officinale).]

Collections and Uses--Late in summer and in fall the milky juice becomes thicker and the bitterness increases and this is the time to collect dandelion root. It should be carefully washed and thoroughly dried. Dandelion roots lose considerably in drying, weighing less than half as much as the fresh roots. The dried root should not be kept too long, as drying diminishes its medicinal activity. It is official in the United States Pharmacopeia.

Dandelion is used as a tonic in diseases of the liver and in dyspepsia.

Imports and Prices--Most of the dandelion root found on the market is collected in central Europe. There has been an unusually large demand for dandelion root during the season of 1907 and according to the weekly records contained in "the Oil, Paint and Drug Reporter" the imports entered at the port of New York from January 1, 1907, to the end of May amounted to about 47,000 pounds. The price ranges from 4 to 10 cents a pound.

Soapwort.

Saponaria Officinalis L.

Other Common Names--Saponaria, saponary, common soapwort, bouncing-bet, soaproot, bruisewort, Boston pink, chimney-pink, crow-soap, hedge-pink, old maid's pink, fuller's-herb, lady-by-the-gate, London-pride, latherwort, mock-gilliflower, scourwort, sheepweed, sweet-betty, wild sweet-william, woods-phlox, world's wonder.

Habitat and Range--By one or another of its many common names this plant, naturalized from Europe, is known almost everywhere, occurring along roadsides and in waste places.

CHAPTER XXXI.

Description of Plant--Soapwort is a rather pretty herbaceous perennial, 1 to 2 feet high, and belonging to the pink family (Silenaceae). Its smooth, stout and erect stem is leafy and sparingly branched, the leaves ovate, 2 to 3 inches long, smooth, prominently ribbed, and pointed at the apex. The bright looking, crowded clusters of pink (or in shady localities whitish) flowers appear from about June until far along in September. The five petals of the corolla are furnished with long "claws" or, in other words, they are narrowly lengthened toward the base and inserted within the tubular and pale green calyx. The seed capsule is oblong and one-celled.

Description of Root--Soapwort spreads by means of its stolons, or underground runners. But the roots, which are rather long are the parts employed in medicine. These are cylindrical, tapering toward the apex, more or less branched, and wrinkled lengthwise. The whitish wood is covered with a brownish red, rather thick bark and the roots break with a short, smooth fracture. It is at first sweetish, bitter, and mucilaginous, followed by a persistently acrid taste, but it has no odor.

[Illustration: Soapwort (Saponaria Officinalis).]

Collection, Prices and Uses--As already indicated, the roots without the runners, should be collected either in spring or autumn. With water they form a lather, like soap, whence the common names soapwort, soaproot, latherwort, etc., are derived. The price ranges from 5 to 10 cents a pound. The roots are employed in medicine for their tonic, alterative and diaphoretic properties. The leaves are also used.

Burdock.

Arctium Lappa L.

Synonym--Lappa major Gaertn.

Pharmacopoeial Name--Lappa.

Other Common Names--Cockle-button, cuckold-dock, beggar's-buttons, hurrbur, stick-buttons, hardock, bardane.

CHAPTER XXXI.

Habitat and Range--Burdock, one of our most common weeds, was introduced from the Old World. It grows along road sides, in fields, pastures and waste places, being very abundant in the Eastern and Central States and in some scattered localities in the West.

Description of Plant--Farmers are only too well acquainted with this coarse, unsightly weed. During the first year of its growth this plant, which is a biennial belonging to the aster family (Asteraceae), produces only a rosette of large, thin leaves from a long, tapering root. In the second year a round, fleshy, and branched stem is produced, the plant when full grown measuring from 3 to 7 feet in height. This stem is branched, grooved, and hairy, bearing very large leaves, the lower ones often measuring 18 inches in length. The leaves are placed alternately on the stem, on long, solid, deeply furrowed leafstalks; they are thin in texture, smooth on the upper surface, pale and woolly underneath; usually heart shaped, but sometimes roundish or oval, with even, wavy, or toothed margins.

The flowers are not produced until the second year, appearing from July until frost. Burdock flowers are purple, in small, clustered heads armed with hooked tips, and the spiny burs thus formed are a great pest, attaching themselves to clothing and to the wool and hair of animals. Burdock is a prolific seed producer, one plant bearing as many as 400,000 seeds.

[Illustration: Burdock (Arctium Lappa), Flowering Branch and Root.]

Description of Rootstock--Burdock has a large, fleshy taproot, which when dry becomes scaly and wrinkled lengthwise and has a blackish brown or grayish brown color on the outside, hard, breaking with a short, somewhat fleshy fracture, and showing the yellowish wood with a whitish spongy center. Sometimes there is a small, white, silky tuft at the top of the root, which is formed by the remains of the bases of the leafstalks. The odor of the root is weak and unpleasant, the taste mucilaginous, sweetish and somewhat bitter. While the root is met with in commerce in its entire state, it is more frequently in broken pieces or in lengthwise slices, the edges of which are turned inward. The roots of other species of Arctium are also employed.

CHAPTER XXXI.

Collection, Prices and Uses--Burdock root is official, and the United States Pharmacopoeia directs that it be collected from plants of the first year's growth, either of Arctium lappa or of other species of Arctium. As Burdock has a rather large, fleshy root, it is difficult to dry and is apt to become moldy, and for this reason it is better to slice the root lengthwise, which will facilitate the drying process. The price ranges from 5 to 10 cents a pound. The best root is said to come from Belgium, where great care is exercised in its collection and curing.

Burdock root is used as an alterative in blood and skin diseases. The seeds and fresh leaves are also used medicinally to a limited extent.

Yellow Dock.

Rumex Crispus L.

Other Common Names--Rumex, curled dock, narrow dock, sour dock.

Habitat and Range--This troublesome weed, introduced from Europe, is now found thruout the United States, occurring in cultivated as well as in waste ground, among rubbish heaps and along the road side.

Description of Plant--Yellow Dock is a perennial plant belonging to the buckwheat family (Polygonaceae), and has a deep, spindle shaped root, from which arises an erect, angular and furrowed stem, attaining a height of from 2 to 4 feet. The stem is branched near the top and leafy, bearing numerous long dense clusters formed by drooping groups of inconspicuous green flowers placed in circles around the stem. The flowers are produced from June to August, and the fruits which follow are in the form of small triangular nuts, like the grain of buckwheat, to which family the dock belongs. So long as the fruits are green and immature they can scarcely be distinguished from the flowers, but as they ripen the clusters take on a rusty brown color. The leaves of the yellow dock are lance shaped, acute, with the margins strongly waved and crisped, the lower long-stalked leaves being blunt or heart shaped at the base from 6 to 8 inches in length, while those nearer the top are narrower and shorter, only 3 to 6 inches in length, short stemmed or stemless.

CHAPTER XXXI.

[Illustration: Yellow Dock (Rumex Crispus), First Year's Growth.]

The broad-leaved dock (Rumex obtusifolius L.), is known also as bitter dock, common dock, blunt-leaved dock, and butter-dock, is a very common weed found in waste places from the New England States to Oregon and south to Florida and Texas. It grows to about the same height as the yellow dock, to which it bears a close resemblance, differing principally in its more robust habit of growth. The stem is stouter than in yellow dock and the leaves, which likewise are wavy along the margin, are much broader and longer. The green flowers appear from June to August and are in rather long, open clusters, the groups rather loose and far apart.

[Illustration: Broad-Leaved Dock (Rumex Obtusifolius), Leaf, Fruiting Spike and Root.]

Description of Roots--Yellow Dock root is large and fleshy, usually from 8 to 12 inches long, tapering or spindle shaped, with few or no rootlets. When dry it is usually twisted and prominently wrinkled, the rather thick, dark, reddish brown bark marked with small scars. The inside of the root is whitish at first, becoming yellowish. The fracture is short, but shows some splintery fibers. The root, as it occurs in commerce, is either entire or occasionally split lengthwise.

The darker colored root of the broad-leaved dock has a number of smaller branches near the crown and more rootlets. Dock roots have but a very faint odor and a bitter, astringent taste.

Collection, Prices and Uses--The roots should be collected in late summer or autumn, after the fruiting tops have turned brown, then washed, either left entire or split lengthwise into halves or quarters and carefully dried. Yellow Dock root ranges from 4 to 6 cents a pound.

In the United States Pharmacopoeia of 1890 "the roots of Rumex crispus and of some other species of Rumex" were official and both of the above-named species are used, but the Yellow Dock (Rumex crispus) is the species most commonly employed in medicine. The docks are largely used for purifying the blood and in the treatment of skin diseases.

CHAPTER XXXI.

The young root leaves of both of the species mentioned are sometimes used in spring as pot herbs.

CHAPTER XXXII.

DRY SOIL PLANTS.

Stillingia.

Stillingia Sylvatica L.

Pharmacopoeial Name--Stillingia.

Other Common Names--Queen's-delight, queen's-root, silverleaf, nettle-potato.

Habitat and Range--This plant is found in dry, sandy soil and in pine barrens from Maryland to Florida west to Kansas and Texas.

Description of Plant--Like most of the other members of the spurge family (Euphorbiaceae), stillingia also contains a milky juice. This indigenous, herbaceous perennial is about 1 to 3 feet in height, bright green and somewhat fleshy, with crowded leaves of a somewhat leathery texture. The leaves are practically stemless and vary greatly in form, from lance shaped, oblong, to oval and elliptical, round toothed or saw toothed. The pale yellow flowers, which appear from April to October, are borne in a dense terminal spike and consist of two kinds, male and female, the male flowers arranged in dense clusters around the upper part of the stalk and the female flowers occurring at the base of the spike. The seeds are contained in a roundish 3-lobed capsule.

Description of Root--Stillingia consists of somewhat cylindrical or slenderly spindle shaped roots from 6 inches to a foot in length, slightly branched, the yellowish white, porous wood covered with a rather thick, reddish brown, wrinkled bark, the whole breaking with a fibrous fracture. As found in commerce, stillingia is usually in short transverse sections, the ends of the sections pinkish and fuzzy with numerous fine, silky bast fibres, and the bark showing scattered yellowish brown resin cells and milk ducts. It has a peculiar unpleasant odor, and a bitter, acrid and pungent taste.

CHAPTER XXXII.

Collection, Prices and Uses--Stillingia root is collected in late autumn or early in spring, usually cut into short, transverse sections and dried. The price ranges from 3 to 5 cents a pound.

This root, which is official in the United States Pharmacopoeia, has been a popular drug in the South for more than a century and is employed principally as an alterative.

American Colombo.

Frasera Carolinensis Walt.

Synonym--Frasera walteri Michx.

Other Common Names--Frasera, meadowpride, pyramid-flower, pyramid-plant, Indian lettuce, yellow gentian, ground-century.

Habit and Range American Colombo occurs in dry soil from the western part of New York to Wisconsin, south to Georgia and Kentucky.

Description of Plant--During the first and second year of the growth of this plant only the root leaves are produced These are generally somewhat rounded at the summit, narrowed toward the base, and larger than the stem leaves, which develop in the third year. The leaves are deep green and produced mostly in whorls of four, the stem leaves being 3 to 6 inches in length and oblong or lance shaped. In the third year the stem is developed and the flowers are produced from June to August. The stem is stout, erect, cylindrical, and 3 to 8 feet in height. The flowers of American Colombo are borne in large terminal, handsome pyramidal clusters, sometimes 2 feet in length, and are greenish yellow or yellowish white, dotted with brown purple. They are slender stemmed, about 1 inch across, with a wheel shaped, 4-parted corolla The seeds are contained in a much compressed capsule. American Colombo is an indigenous perennial and belongs to the gentian family (Gentianaceae.)

Description of Root--The root is long, horizontal, spindle shaped, yellow, and wrinkled. In the fresh state it is fleshy and quite heavy. The

CHAPTER XXXII.

American Colombo root of commerce, formerly in transverse slices, now generally occurs in lengthwise slices. The outside is yellowish or pale orange and the inside spongy and pale yellow. The taste is bitter. American Colombo root resembles the official gentian root in taste and odor, and the uses are also similar.

[Illustration: American Colombo (Frasera Carolinensis), Leaves, Flowers and Seed Pods.]

Collection, Prices and Uses--The proper time for collecting American Colombo root is in the autumn of the second year or in March or April of the third year. It is generally cut into lengthwise slices before drying. The price of American Colombo root ranges from 3 to 5 cents a pound.

The dried root, which was official in the United States Pharmacopoeia from 1820 to 1880, is used as a simple tonic. In the fresh state the root possesses emetic and cathartic properties.

Couch-Grass.

Agropyron repens (L.) Beauv.

Synonym--Triticum repens L.

Pharmacopoeial Name--Triticum.

Other Common Names--Dog-grass, quick-grass, quack-grass, quitch-grass, quake-grass, scutch-grass, twitch-grass, witch-grass, wheat-grass, creeping wheat-grass, devil's grass, durfa-grass, Durfee-grass, Dutch-grass, Fin's-grass, Chandler's-grass.

Habitat and Range--Like many of our weeds, couch-grass was introduced from Europe, and is now one of the worst pests the farmer has to contend with, taking possession of the cultivated ground and crowding out valuable crops. It occurs most abundantly from Maine to Maryland, westward to Minnesota and Minnesota, and is spreading on farms on the Pacific slope, but is rather sparingly distributed in the South.

CHAPTER XXXII.

[Illustration: Couch-Grass (Agrophyron Repens).]

Description of Plant--Couch-grass is rather coarse, 1 to 3 feet high, and when in flower very much resemble rye or beardless wheat. Several round, smooth, hollow stems, thickened at the joints, are produced from the long, creeping, jointed rootstock. The stems bear 5 to 7 leaves from 3 to 12 inches long, rough on the upper surface and smooth beneath, while the long, cleft leaf sheaths are smooth. The solitary terminal flowering heads or spikes are compressed, and consist of two rows of spikelets on a wavy and flattened axis. These heads are produced from July to September. Couch Grass belongs to the grass family (Poaceae.)

Description of Rootstock--The pale yellow, smooth rootstock is long, tough and jointed, creeping along underneath the ground, and pushing in every direction. As found in the stores, it consists of short, angular pieces, from one eighth to one-fourth of an inch long, of a shining straw color, and hollow. These pieces are odorless, but have a somewhat sweetish taste.

Collection, Prices and Uses--Couch-Grass, which is official in the United States Pharmacopoeia, should be collected in spring, carefully cleaned, and the rootlets removed. The rootstock (not rootlets) is then cut into short pieces about two-fifths of an inch in length, for which purpose an ordinary feed-cutting machine may be used, and thoroughly dried.

Couch-Grass is usually destroyed by plowing up and burn ing, for if any of the joints are permitted to remain in the soil new plants will be produced. But, instead of burning, the rootstocks may be saved and prepared for the drug market in the manner above stated. The prices range from 3 to 5 cents a pound. At present Couch-Grass is collected chiefly in Europe.

A fluid extract is prepared from Couch-Grass, which is used in affections of the kidney and bladder.

Echinacea.

CHAPTER XXXII.

Brauneria Angustifolia (DC) Heller.

Synonym--Echinacea angustifolia DC.

Other Common Names--Pale-purple coneflower, Sampson-root, niggerhead (in Kansas.)

Habitat and Range--Echinacea is found in scattered patches in rich prairie soil or sandy soil from Alabama to Texas and northwestward, being most abundant in Kansas and Nebraska. Tho not growing wild in the Eastern States, It has succeeded well under cultivation in the testing gardens of the Department of Agriculture at Washington, D. C.

[Illustration: Echinacea (Brauneria Angustifolia).]

Description of Plant--This native herbaceous perennial, belonging to the aster family (Asteraceae), grows to a height of from 2 to 3 feet. It sends up a rather stout bristly-hairy stem, bearing thick rough-hairy leaves, which are broadly lance shaped or linear lance shaped, entire, 3 to 8 inches long, narrowed at each end, and strongly three nerved. The lower leaves have slender stems, but as they approach the top of the plant the stems become shorter and some of the upper leaves are stemless.

The flower heads appearing from July to October, are very pretty, and the plant would do well as an ornamental in gardens. The flowers remain on the plant for a long time, and the color varies from whitish rose to pale purple. The head consists of ray flowers and disk flowers, the former constituting the "petals" surrounding the disk, and the disk itself being composed of small, tubular, greenish yellow flowers. When the flowers first appear the disk is flattened or really concave, but as the flowering progresses it becomes conical in shape. The brown fruiting heads are conical, chaffy, stiff and wiry.

Description of Root--Echinacea has a thick, blackish root, which in commerce occurs in cylindrical pieces of varying length and thickness. The dried root is grayish brown on the outside, the bark wrinkled lengthwise and sometimes spirally twisted. It breaks with a short, weak fracture, showing yellow or greenish yellow wood edges, which give

CHAPTER XXXII.

the impression that the wood is decayed.

The odor is scarcely perceptible and the taste is mildly aromatic, afterwards becoming acrid and inducing a flow of saliva.

Collection, Prices and Uses--The root of Echinacea is collected in autumn and brings from 20 to 30 cents a pound. It is said that Echinacea varies greatly in quality due chiefly to the locality in which it grows. According to J. U. Lloyd, the best quality comes from the prairie lands of Nebraska and that from marshy places is inferior.

Echinacea is said to be an alterative and to promote perspiration and induce a flow of saliva. The Indians used the freshly scraped roots for the cure of snake bites.

Aletris.

Aletris Farinosa L.

Other Common Names--Stargrass, blazingstar, mealy starwort, starwort, unicorn-root, true unicorn-root, unicorn-plant, unicorn's-horn, colic-root, devil's-bit, ague-grass, ague-root, aloe-root, crow-corn, huskwort.

A glance at these common names will show many that have been applied to other plants, especially to Chamaelirium, with which Aletris is so much confused. In order to guard against this confusion as much as possible, it is best not to use the common names of this plant at all, referring to it only by its generic name, Aletris.

[Illustration: Aletris (Aletris Farinosa).]

Habitat and Range--Aletris occurs in dry, generally sandy soil, from Maine to Minnesota, Florida and Tennessee.

Description of Plant--As stated under Chamaelirium, this plant is often confused with the former by collectors and others, although there seems to be no good reason why this should be so. The plants do not resemble each other except in habit of growth, and the trouble

CHAPTER XXXII.

undoubtedly arose from a confusion of the somewhat similar common names of the plants, as, for instance, "stargrass" and "starwort."

Aletris may be at once distinguished by the grasslike leaves, which spread out on the ground in the form of a star, and by the slender spikes of rough, mealy flowers.

This native perennial, belonging to the lily family (Liliaceae), is an erect, slender herb, 1 1/2 to 3 feet tall, with basal leaves only. These leaves are grasslike, from 2 to 6 inches long, and have a yellowish green or willow-green color. As already stated, they surround the base of the stem in the form of a star. Instead of stem leaves, there are very small, leaflike bracts placed at some distance apart on the stem. From May to July the erect flowering spike, from 4 to 12 inches long, is produced, bearing white, urn-shaped flowers, sometimes tinged with yellow at the apex, and having a rough, wrinkled and mealy appearance. The seed capsule is ovoid, opening by three halves, and containing many seeds. When the flowers in the spike are still in bud, there is a suggestion of resemblance to the female spike of Chamaelirium with its fruit half formed.

Several other species are recognized by botanists, namely, Aletris Aurea Walt., A. lutea Small, and A. obovata Nash, but aside from the flowers, which in aurea and lutea are yellow, and slight variations in form, such as a more contracted perianth, the differences are not so pronounced that the plants would require a detailed description here. They have undoubtedly been collected with Aletris farinosa for years, and are sufficiently like it to be readily recognized.

Description of Rootstock--Not only have the plants of Aletris and Chamaelirium been confused, but the rootstocks as well. There is, however, no resemblance between them.

Aletris has a horizontal rootstock from one-half to 1 1/2 inches in length, rough and scaly, and almost completely hidden by the fibrous roots and remains of the basal leaves. Upon close examination the scars of former leaf stems may be seen along the upper surface. The rootlets are from 2 to 10 inches in length, those of recent growth whitish and covered with several layers of epidermis which gradually

CHAPTER XXXII.

peel off, and the older rootlets of the rootstock showing this epidermis already scaled off, leaving only the hard, brown, woody center. The rootstock in commerce almost invariably shows at one end a tuft of the remains of the basal leaves, which do not lose their green color. It is grayish brown outside, whitish within, and breaks with a mealy fracture. It has no odor, and a starchy taste, followed by some acridity, but no bitterness.

Collection, Prices and Uses--Aletris should be collected in autumn, and there is no reason why collectors should make the common mistake of confusing Aletris with Chamaelirium. By comparing the description of Aletris with that of Chamaelirium, it will be seen that there is scarcely any resemblance. Aletris ranges from 30 to 40 cents a pound.

As indicated under Chamaelirium, the medicinal properties have also been considered the same in both plants, but Aletris is now regarded of value chiefly in digestive troubles. Aletris was official in the United States Pharmacopoeia from 1820 to 1870.

Wild Indigo.

Baptisia Tinctoria (L.) R. Br.

Other Common Names--Baptisia, indigo-weed, yellow indigo, American indigo, yellow broom, indigo-broom, clover-broom, broom-clover, horsefly-weed, shoofly, rattlebush.

Habitat and Range--This native herb grows on dry, poor land, and is found from Maine to Minnesota, south to Florida and Louisiana.

Description of Plant--Many who have been brought up in the country will recognize in the wild indigo the plant so frequently used by farmers, especially in Virginia and Maryland, to keep flies away from horses, bunches of it being fastened to the harness for this purpose.

[Illustration: Wild Indigo (Baptisia Tinctoris) Branch Showing Flowers and Seed Pods.]

CHAPTER XXXII.

Wild Indigo grows about 2 to 3 feet in height and the clover-like blossoms and leaves will show at once that it belongs to the same family as the common clover, namely, the pea family (Fabaceae). It is an erect, much-branched, very leafy plant of compact growth, the 3-leaved, bluish green foliage somewhat resembling clover leaves. The flowers, as already stated, are like common clover flowers--that is, not like clover heads, but the single flowers composing these; they are bright yellow, about one-half inch in length and are produced in numerous clusters which appear from June to September. The seed pods, on stalks longer than the calyx, are nearly globular or ovoid and are tipped with an awl shaped style.

Another species, said to possess properties similar to those of baptisia tinctoria and substituted for it, is B. alba R. Br., called the white wild indigo. This plant has white flowers and is found in the Southern States and on the plains of the Western States.

Description of Root--Wild Indigo has a thick, knotty crown or head, with several stem scars, and a round, fleshy root, sending out cylindrical branches and rootlets almost 2 feet in length. The white woody interior is covered with a thick, dark brown bark, rather scaly or dotted with small, wart-like excrescences. The root breaks with a tough, fibrous fracture. There is a scarcely perceptible odor and the taste, which resides chiefly in the bark, is nauseous, bitter and acrid.

Collection, Prices and Uses--The root of Wild Indigo is collected in autumn, and brings from 4 to 8 cents a pound.

Large doses of Wild Indigo are emetic and cathartic and may prove dangerous. It also has stimulant, astringent and antiseptic properties, and is used as a local application to sores, ulcers, etc.

The herb is sometimes employed like the root and the entire plant was official from 1830 to 1840.

In some sections the young, tender shoots are used for greens, like those of pokeweed, but great care must be exercised to gather them before they are too far advanced in growth, as otherwise bad results will follow.

CHAPTER XXXII.

A blue coloring matter has been prepared from the plant and used as a substitute for indigo, to which, however, it is very much inferior.

Pleurisy-Root.

Asclepias Tuberosa L.

Pharmacopoeial Name--Asclepias.

Other Common Names--Butterfly weed, Canada-root, Indian-posy, orange-root, orange swallowwort, tuberroot, whiteroot, windroot, yellow or orange milkweed.

Habitat and Range--Pleurisy-Root flourishes in the open or in the pine woods, in dry, sandy or gravelly soil, usually along the banks of streams. Its range extends from Ontario and Maine to Minnesota, south to Florida, Texas and Arizona, but it is found in greatest abundance in the South.

Description of Plant--This is a very showy and ornamental perennial plant, indigenous to this country, and belonging to the milkweed family (Asclepiadaceae); it is erect and rather stiff in habit, but with brilliant heads of bright orange-colored flowers that attract attention from afar.

The stems are rather stout, erect, hairy, about 1 to 2 feet in height, sometimes branched near the top, and bearing a thick growth of leaves. These are either stemless or borne on short stems, are somewhat rough to the touch, 2 to 6 inches long, lance shaped or oblong, the apex either sharp pointed or blunt, with a narrow, rounded or heart shaped base. The flower heads, borne at the ends of the stem and branches, consist of numerous, oddly shaped orange colored flowers. The corolla is composed of five segments, which are reflexed or turned back and the crown has five erect or spreading "hoods," within each of which is a slender incurved horn. The plant is in flower for some time, usually from June to September, followed late in the fall by pods, which are from 4 to 5 inches long, green, tinged with red, finely hairy on the outside, and containing the seeds with their long, silky hairs. Unlike the other milkweeds, the Pleurisy Root contains little or no milky juice.

Description of Root--The root of this plant is large, white and fleshy, spindle shaped, branching. As found in commerce it consists of lengthwise or crosswise pieces from 1 to 6 inches in length and about three-fourths of an inch in thickness. It is wrinkled lengthwise and also transversely and has a knotty head. The thin bark is orange brown and the wood yellowish, with white rays. It has no odor and a somewhat bitter, acrid taste.

[Illustration: Pleurisy-Root (Asclepias Tuberosa).]

Collection, Prices and Uses--The root, which is usually found rather deep in the soil, is collected in autumn, cut into transverse or lengthwise slices and dried. The price ranges from 6 to 10 cents a pound.

Pleurisy-Root was much esteemed by the Indians, has long been used in domestic practice, and is official in the United States Pharmacopoeia. It is used in disordered digestion and in affections of the lungs, in the last-named instance to promote expectoration, relieve pains in the chest, and induce easier breathing. It is also useful in producing perspiration.

Other Species--Besides the official Pleurisy-Root there are two other species of Asclepias which are employed to some extent for the same purposes, namely, the common milkweed and the swamp-milkweed.

The common milkweed (Asclepias syriaca L.) is a perennial, native in fields and waste places from Canada to North Carolina and Kansas. It has a stout, usually simple stem 3 to 5 feet in height and oblong or oval leaves, smooth on the upper surface and densely hairy beneath. The flowers, similar in form to those of Asclepias tuberosa, are pinkish purple and appear from June to August, followed by erect pods 3 to 5 inches long, woolly with matted hair and covered with prickles and borne on recurved stems. The plant contains an abundance of milky juice.

The root of the common milkweed is from 1 to 6 feet long, cylindrical and finely wrinkled. The short branches and scars left by former stems give the root a round, knotty appearance. The bark is thick, grayish brown and the inside white, the root breaking with a short, splintery fracture. Common milkweed root has a very bitter taste, but no odor.

It is collected in autumn and cut into transverse slices before drying. Common milkweed ranges from 6 to 8 cents a pound.

Swamp-milkweed (Asclepias incarnata L.) is a native perennial herb found in swamps from Canada to Tennessee and Kansas. The slender stem, leafy to the top, is 1 to 2 feet in height, branched above, the leaves lance shaped or oblong lance shaped. The flowers, also similar to those A tuberosa, appear from July to September, and are flesh colored or rose colored. The pods are 2 to 3 1/2 inches long, erect, and very sparingly hairy.

The root of the swamp-milkweed, which is also collected in autumn, is not quite an inch in length, hard and knotty, with several light brown rootlets. The tough white wood, which has a thick, central pith, is covered with a thin, yellowish brown bark. It is practically without odor, and the taste, sweetish at first, finally becomes bitter. This root brings about 3 cents a pound.

CHAPTER XXXIII.

RICH SOIL PLANTS.

Bloodroot.

Sanguinaria Canadensis L.

Pharmacopoeial--Sanguinaria.

Other Common Names--Redroot, red puccoon, red Indian-paint, puccoon-root, coonroot, white puccoon, pauson, snakebite, sweet-slumber, tetterwort, tumeric.

Habitat and Range--Bloodroot is found in rich, open woods from Canada south to Florida and west to Arkansas and Nebraska.

Description of Plant--This indigenous plant is among the earliest of our spring flowers, the waxy-white blossom, enfolded by the grayish green leaf, usually making its appearance early in April. The stem and root contain a blood-red juice. Bloodroot is a perennial and belongs to the same family as the opium poppy, the Papaveraceae. Each bud on the thick, horizontal rootstock produces but a single leaf and a flowering scape, reaching about 6 inches in height. The plant is smooth and both stem and leaves, especially when young, present a grayish green appearance, being covered with a "bloom" such as is found on some fruits. The leaves are palmately 5 to 9 lobed, the lobes either cleft at the apex or having a wavy margin, and are borne on leaf stems about 6 to 14 inches long. After the plants have ceased flowering the leaves, at first only 3 inches long and 4 to 5 inches broad, continue to expand until they are about 4 to 7 inches long and 6 to 12 inches broad. The under side of the leaf is paler than the upper side and shows prominent veins. The flower measures about 1 inch across, is white, rather waxlike in appearance, with numerous golden-yellow stamens in the center. The petals soon fall off, and the oblong, narrow seed pod develops, attaining a length of about an inch.

[Illustration: Bloodroot (Sanguinaria Canadensis) Flowering Plant with Rootstock.]

CHAPTER XXXIII.

Description of Rootstock--When dug out of the ground Bloodroot is rather thick, round and fleshy, slightly curved at the ends, and contains a quantity of blood-red juice. It is from 1 to 4 inches in length, from one-half to 1 inch in thickness, externally reddish brown, internally a bright red blood color, and produces many thick, orange colored rootlets.

The rootstock shrinks considerably in drying, the outside turning dark brown and the inside orange-red or yellowish with numerous small red dots, and it breaks with a short, sharp fracture. It has but a slight odor and the taste is bitter and acrid and very persistent. The powdered root causes sneezing.

Collection, Prices and Use--The rootstock should be collected in autumn, after the leaves have died, and after curing, it should be stored in a dry place, as it rapidly deteriorates if allowed to become moist. Age also impairs its acridity. The price paid to collectors for this root ranges from about 5 to 10 cents per pound.

Bloodroot was well known to the American Indians, who used the red juice as a dye for skins and baskets and for painting their faces and bodies. It is official in the United States Pharmacopoeia and is used as a tonic, alterative, stimulant and emetic.

Pinkroot.

Spigelia Marilandica L.

Pharmacopoeial Name--Spigelia.

Other Common Names--Carolina pinkroot, pinkroot, Carolina pink, Maryland pink, Indian pink, starbloom, wormgrass, wormweed, American wormroot.

Habitat and Range--This pretty little plant is found in rich woods from New Jersey to Florida, west to Texas and Wisconsin, but occurring principally in the Southern States. It is fast disappearing, however from its native haunts.

CHAPTER XXXIII.

[Illustration: Pinkroot (Spigelia Marilandica).]

Description of Plant--Pinkroot belongs to the same family as the yellow jasmine, namely, the Logania family (Loganiaceae), noted for its poisonous species. It is a native perennial herb, with simple, erect stem 6 inches to 1 1/2 feet high, nearly smooth. The leaves are stemless, generally ovate, pointed at the apex and rounded or narrowed at the base; they are from 2 to 4 inches long, one-half to 2 inches wide, smooth on the upper surface, and only slightly hairy on the veins on the lower surface. The rather showy flowers are produced from May to July in a terminal one-sided spike; they are from 1 to 2 inches in length, somewhat tube shaped, narrowed below, slightly inflated toward the center, and again narrowed or contracted toward the top, terminating in five lance shaped lobes; the flowers are very showy, with their brilliant coloring--bright scarlet on the outside, and the inside of the tube, and the lobes a bright yellow. The seed capsule is double, consisting of two globular portions more or less united, and containing numerous seeds.

Description of Rootstock--The rootstock is rather small, from 1 to 2 inches in length and about one-sixteenth of an inch in thickness. It is somewhat crooked or bent, dark brown, with a roughened appearance of the upper surface caused by cup shaped scars, the remains of former annual stems. The lower surface and the sides have numerous long, finely branched, lighter colored roots, which are rather brittle. Pinkroot has a pleasant, aromatic odor, and the taste is described as sweetish, bitter and pungent.

Collection, Prices and Uses--Pinkroot is collected after the flowering period. It is said to be scarce, and was reported as becoming scarce as long ago as 1830. The price paid to collectors ranges from 25 to 40 cents a pound.

The roots of other plants, notably those of the East Tennessee pinkroot (Ruellia ciliosa Pursh), are often found mixed with the true Pinkroot, and the Ruellia ciliosa is even substituted for it. This adulteration or substitution probably accounts for the inertness which has sometimes been attributed to the true Pinkroot and which has caused it to fall into more or less disuse. It has long been known that

CHAPTER XXXIII.

the true Pinkroot was adulterated, but this adulteration was supposed to be caused by the admixture of Carolina phlox (Phlox Carolina L., now known as Phlox ovata L.), but this is said now to be no part of the substitution.

The rootstock of Ruellia ciliosa is larger and not as dark as that of the Maryland pinkroot and has fewer and coarser roots, from which the bark readily separates, leaving the whitish wood exposed.

Pinkroot was long known by the Indians, and its properties were made known to physicians by them. It is official in the United States Pharmacopoeia and is used principally as an anthelmintic.

Indian-Physic.

Porteranthus Trifoliatus (L.) Britton.

Synonym--Gilenia Trifoliata Moench.

Other Common Names--Gilenia, bowman's-root, false ipecac, western dropwort, Indian-hippo.

Habitat and Range--Indian-Physic is native in rich woods from New York to Michigan, south to Georgia and Missouri.

Description of Plant--The reddish stems of this slender, graceful perennial of the rose family (Rosaceae) are about 2 to 3 feet high, several erect and branched stems being produced from the same root. The leaves are almost stemless and trifoliate; that is, composed of three leaflets. They are ovate or lanceolate, 2 to 3 inches long, narrowed at the base, smooth and toothed. The nodding, white pinkish flowers are few, produced in loose terminal clusters from May to July. The five petals are long, narrowed or tapering toward the base, white or pinkish, and inserted in the tubular, somewhat bell shaped, red tinged calyx. The seed pods are slightly hairy.

At the base of the leaf stems are small leaflike parts, called stipules, which in this species are very small, linear and entire. In the following species, which is very similar to trifoliatus and collected with it, the stipules, however, are so much larger that they form a prominent character, which has given rise to its specific name, stipulatus.

[Illustration: Indian Physic (Porteranthus Trifoliatus).]

Porteranthus stipulatus (Muhl.) Britton (Syn. Gillenia stipulacea Nutt.) is found in similar situations as P. trifoliatus, but generally farther west, its range extending from western New York to Indiana and Kansas, south to Alabama, Louisiana and Indian Territory. The general appearance of this plant is very similar to that of P. trifoliatus. It grows to about the same height, but is generally more hairy, the leaflets narrower and more deeply toothed, and the flowers perhaps a trifle smaller. The stipules, however, will generally serve to distinguish it. These are large, broad, ovate, acute at the apex, sharply and deeply notched and so much like leaves that but for their position at the base of the leaf stems they might easily be mistaken for them.

With the exception of the name American ipecac applied to this plant, the common names of Porteranthus trifoliatus are also used for P. stipulatus. The roots of both species are collected and used for the same purpose.

Description of Roots--The root Porteranthus trifoliatus is thick and knotty, with many smoothish, reddish brown rootlets, the latter in drying becoming wrinkled lengthwise and showing a few transverse fissures or breaks in the bark, and the interior white and woody. There is practically no odor and the woody portion is tasteless, but the bark, which is readily separable, is bitter, increasing the flow of saliva.

Porteranthus stipulatus has a larger, more knotty root, with rootlets that are more wavy, constricted or marked with numerous transverse rings, and the bark fissured or breaking from the white woody portion at frequent intervals.

Collection, Prices and Uses--The roots of both species are collected in autumn. The prices range from 2 to 4 cents a pound.

Indian-Physic or bowman's root, as these names imply, was a popular remedy with the Indians, who used it as an emetic. From them the

CHAPTER XXXIII.

white settlers learned of its properties and it is still used for its emetic action. This drug was at one time official in the United States Pharmacopoeia, from 1820 to 1880. Its action is said to resemble that of ipecac.

Wild Sarsaparilla.

Arala Nudicaulis L.

Other Common Names--False sarsaparilla, Virginia sarsaparilla, American sarsaparilla, small spikenard, rabbit's-root, shotbush, wild licorice.

Habitat and Range--Wild Sarsaparilla grows in rich, moist woods from Newfoundland west to Manitoba and south to North Carolina and Missouri.

Description of Plant--This native herbaceous perennial, belonging to the ginseng family (Araliaceae), produces a single, long-stalked leaf and flowering stalk from a very short stem, both surrounded or sheathed at the base by thin, dry scales. The leafstalk is about 12 inches long divided at the top into three parts, each division bearing five oval, toothed leaflets from 2 to 5 inches long, the veins on the lower surface sometimes hairy.

The naked flowering stalk bears three spreading clusters of small, greenish flowers, each cluster consisting of from 12 to 30 flowers produced from May to June, followed later in the season by purplish black roundish berries, about the size of the common elderberries.

[Illustration: Wild Sarsaparilla (Aralia Nudicaulis).]

Description of Rootstock--Wild Sarsaparilla rootstock has a very fragrant, aromatic odor. Rabbits are said to be very fond of it, whence one of the common names, "rabbit's-root," is derived. The rootstock is rather long, horizontally creeping, somewhat twisted, and yellowish brown on the outside. The taste is warm and aromatic. The dried rootstock is brownish, gray and wrinkled lengthwise on the outside, about one-fourth of an inch in thickness, the inside whitish with a spongy pith. The taste is sweetish and somewhat aromatic.

Collection, Prices and Uses--The root of Wild Sarsaparilla is collected in autumn, and brings from 5 to 8 cents a pound.

This has long been a popular remedy, both among the Indians and domestic practice, and was official in the United States Pharmacopoeia from 1820 to 1880. Its use is that of an alterative, stimulant and diaphoretic and in this it resembles the official sarsaparilla obtained from tropical America.

Similar Species--The American spikehead (Aralia racemosa L.), known also as spignet, spiceberry, Indian-root, petty-morrel, life-of-man and old-man's-root, is employed like Aralia nudicaulis. It is distinguished from this by its taller, herbaceous habit, its much-branched stem from 3 to 6 feet high and very large leaves consisting of thin, oval, heart shaped, double saw-toothed leaflets. The small, greenish flowers are arranged in numerous clusters, instead of only three as in nudicaulis and also appear somewhat later, namely, from July to August. The berries are roundish, reddish brown, or dark purple.

The rootstock is shorter than that of nudicaulis and much thicker, with prominent stem scars, and furnished with numerous, very long, rather thin roots. The odor and taste are stronger than in nudicaulis. It is also collected in autumn, and brings from 4 to 8 cents a pound.

The American spikenard occurs in similar situations as nudicaulis, but its range extends somewhat farther South, Georgia being given as the Southern limit.

The California spikenard (Aralia californica Wats.) may be used for the same purpose as the other species. The plant is larger than Aralia racemosa, but otherwise is very much like it. The root is also larger than that of A. racemosa.

CHAPTER XXXIV.

MEDICINAL HERBS.

American Angelica.

Angelica Atropurpurea L.

Synonym--Archangelica atropurpurea Hoffn.

Other Common Names--Angelica, purple-stemmed angelica, great angelica, high angelicam, purple angelica, masterwort.

Habitat and Range--American Angelica is a native herb, common in swamps and damp places from Labrador to Delaware and west to Minnesota.

Description of Plant--This strong-scented, tall, stout perennial reaches a height of from 4 to 6 feet, with a smooth, dark purple, hollow stem 1 to 2 inches in diameter. The leaves are divided into three parts, each of which is again divided into threes; the rather thin segments are oval or ovate, somewhat acute, sharply toothed and sometimes deeply cut, and about 2 inches long. The lower leaves sometimes measure 2 feet in width, while the upper ones are smaller, but all have very broad, expanded stalks. The greenish white flowers are produced from June to July in somewhat roundish, many-rayed umbels or heads, which sometimes are 8 to 10 inches in diameter. The fruits are smooth, compressed and broadly oval. American Angelica root is branched, from 3 to 6 inches long, and less than an inch in diameter. The outside is light, brownish gray, with deep furrows, and the inside nearly white, the whole breaking with a short fracture and the thick bark showing fine resin dots. It has an aromatic odor, and the taste at first is sweetish and spicy, afterwards bitter. The fresh root is said to possess poisonous properties.

The root of the European or garden angelica (Angelica officinalis Moench) supplies much of the angelica root of commerce. This is native in northern Europe and is very widely cultivated, especially in Germany, for the root.

CHAPTER XXXIV.

[Illustration: American Angelica (Angelica Atropurpurea).]

Collection, Prices and Uses--The root is dug in autumn and carefully dried. Care is also necessary in preserving the root, as it is very liable to the attacks of insects. American Angelica root ranges from 6 to 10 cents a pound.

American Angelica root, which was official in the United States Pharmacopoeia from 1820 to 1860 is used as an aromatic, tonic, stimulant, carminative, diuretic and diaphoretic. In large doses it acts as an emetic.

The seeds are also employed medicinally.

Comfrey.

Symphytum Officinale L.

Other Common Names--Symphytum, healing herb, knitback, ass-ear, backwort, blackwort, bruisewort, gum-plant, slippery-root.

Habitat and Range--Comfrey is naturalized from Europe and occurs in waste places from Newfoundland to Minnesota, south to Maryland.

[Illustration: Comfrey (Symphytum Officinale).]

Description of Plant--This coarse, rough, hairy, perennial herb is from 2 to 3 feet high, erect and branched, with thick, rough leaves, the lower ones ovate lance shaped, 3 to 10 inches long, pointed at the apex, and narrowed at the base into margined stems. The uppermost leaves are lance shaped, smaller and stemless. Comfrey is in flower from June to August, the purplish or dirty white, tubular, bell shaped flowers numerous and borne in dense terminal clusters. The nutlets which follow are brown, shinning and somewhat wrinkled. Comfrey belongs to the borage family (Boraginaceae.)

Description of Root--Comfrey has a large, deep, spindle-shaped root, thick and fleshy at the top, white inside and covered with a thin, blackish brown bark. The dried root is hard, black and very deeply and roughly wrinkled, breaking with a smooth, white, waxy fracture. As it occurs in commerce it is in pieces ranging from about an inch to several inches in length, only about one-fourth of an inch in thickness, and usually considerably bent. It has a very mucilaginous, somewhat sweetish and astringent taste, but no odor.

Collection, Prices and Uses--The root is dug in autumn, or sometimes in early spring. Comfrey root when first dug is very fleshy and juicy, but about four-fifths of its weight is lost in drying. The price ranges from 4 to 8 cents a pound.

The mucilaginous character of Comfrey root renders it useful in coughs and diarrheal complaints. Its action is demulcent and slightly astringent.

The leaves are also used to some extent.

Elecampane.

Inula Helenium L.

Other Common Names--Inula, inul, horseheal, elf-dock, elfwort, horse-elder, scabwort, yellow starwort, velvet dock, wild sunflower.

Habitat and Range--This perennial herb has been naturalized from Europe, and is found along the roadsides and in fields and damp pastures from Nova Scotia to North Carolina, westward to Missouri and Minnesota. It is a native also in Asia.

Description of Plant--When in flower elecampane resembles the sunflower on a small scale. Like the sunflower, it is a member of the aster family (Asteraceae). It is a rough plant, growing from 3 to 6 feet in height, but producing during the first year only root leaves, which attain considerable size. In the following season the stout densely hairy stem develops, attaining a height of from 3 to 6 feet.

[Illustration: Elecampane (Inula Helenium).]

CHAPTER XXXIV.

The leaves are broadly oblong in form, toothed, the upper surface rough and the under side densely soft-hairy. The basal or root leaves are borne on long stems, and are from 10 to 20 inches long and 4 to 8 inches wide, while the upper leaves are smaller and stemless or clasping.

About July to September the terminal flower heads are produced, either singly or a few together. As already stated, these flower heads look very much like small sunflowers, 2 to 4 inches broad, and consist of long, narrow, yellow rays, 3 toothed at the apex, and the disk also is yellow.

Description of Root--Elecampane has a large, long, branching root, pale yellow on the outside and whitish and fleshy within. When dry the outside turns a grayish brown or dark brown, and is generally finely wrinkled lengthwise. As found in commerce, elecampane is usually in transverse or lengthwise slices, light yellow or grayish and fleshy internally, dotted with numerous shining resin cells, and with overlapping brown or wrinkled bark. These slices become flexible in damp weather and tough, but when they are dry they break with a short fracture. The root has at first a strongly aromatic odor, which has been described by some as resembling a violet odor, but this diminished in drying. The taste is aromatic, bitterish and pungent.

Collection, Prices, and Uses--The best time for collecting elecampane is in the fall of the second year. If collected later than that the roots are apt to be stringy and woody. Owing to the interlacing habit of the rootlets, much dirt adheres to the root, but it should be well cleaned, cut into transverse or lengthwise slices, and carefully dried in the shade. Collectors receive from 3 to 5 cents a pound for this root.

Elecampane, which was official in the United States Pharmacopeia of 1890, is much used in affections of the respiratory organs, in digestive and liver disorders, catarrhal discharges and skin diseases.

Queen-of-the-Meadow.

Eupatorium Purpureum.

CHAPTER XXXIV.

Other Common Names--Gravelroot, Indian gravelroot, joe-pye-weed, purple boneset, tall boneset, kidney root, king-of-the-meadow, marsh-milkweed, motherwort, niggerweed, quillwort, slunkweed, trumpetweed.

Habitat and Range--This common native perennial herb occurs in low grounds and dry woods and meadows from Canada to Florida and Texas.

Description of Plant--The stout, erect, green or purple stem of this plant grows from 3 to 10 feet in height and is usually smooth, simple or branched at the top. The thin, veiny leaves are 4 to 12 inches long, 1 to 3 inches wide, ovate or ovate lance shaped, sharp pointed, toothed and placed around the stem in whorls of three to six. While the upper surface of the leaves is smooth, there is usually a slight hairiness along the veins on the lower surface, otherwise smooth. Toward the latter part of the summer and in early fall queen-of-the-meadow is in flower, producing 5 to 15 flowered pink or purplish heads, all aggregated in large compound clusters which present a rather showy appearance. This plant belongs to the aster family (Asteraceae).

[Illustration: Queen-of-the-Meadow (Eupatorium Purpureum).]

Another species which is collected with this and for similar purposes, and by some regarded as only a variety, is the spotted boneset or spotted joe-pye-weed (Eupatorium maculatum L.) This is very similar to E. purpureum, but it does not grow so tall, is rough-hairy and has the stem spotted with purple. The thicker leaves are coarsely toothed and in whorls of three to five and the flower clusters are flattened at the top rather than elongated as in E. purpureum.

It is found in moist soil from New York to Kentucky, westward to Kansas, New Mexico, Minnesota, and as far up as British Columbia.

Description of Root--Queen-of-meadow root, as it occurs in commerce, is blackish and woody, furnished with numerous long dark-brown fibers, which are furrowed or wrinkled lengthwise and whitish within. It has a bitter, aromatic and astringent taste.

CHAPTER XXXIV.

Collection, Prices and Uses--The root is collected in autumn and is used for its astringent and diuretic properties. It was official in the United States Pharmacopeia from 1820 to 1840. The price ranges from 2 1/2 to 4 cents a pound.

CHAPTER XXXV.

MEDICINAL SHRUBS.

Hydrangea.

Hydrangea Arborescens L.

Other Common Names--Wild hydrangea, seven-barks.

Habitat and Range--Hydrangea frequents rocky river banks and ravines from the southern part of New York to Florida, and westward to Iowa and Missouri, being especially abundant in the valley of the Delaware and southward.

Description of Plant--Hydrangea is an indigenous shrub, 5 to 6 feet or more in height, with weak twigs, slender leaf stems and thin leaves. It belongs to the hydrangea family (Hydrangeaceae). The leaves are oval or sometimes heart shaped, 3 to 6 inches long, sharply toothed, green on both sides, the upper smooth and the lower sometimes hairy. The shrub is in flower from June to July, producing loose, branching terminal heads of small, greenish white flowers, followed by membranous, usually 2-celled capsules, which contain numerous seeds. Sometimes hydrangea will flower a second time early in fall.

A peculiar characteristic of this shrub and one that has given rise to the common name "seven-barks", is the peeling off of the stem bark, which comes off in several successive layers of thin, different colored bark.

Description of Root--The root is roughly branched and when first taken from the ground is very juicy, but after drying it becomes hard. The smooth white and tough wood is covered with a thin, pale-yellow or light-brown bark, which readily scales off. The wood is tasteless, but the bark has a pleasant aromatic taste, becoming somewhat pungent.

[Illustration: Hydrangea (Hydrangea Arborescens).]

Collection, Prices and Uses--Hydrangea root is collected in autumn and as it becomes very tough after drying and difficult to bruise it is best to cut the root in short transverse pieces while it is fresh and still juicy and dry it in this way. The price ranges from 2 to 7 cents a pound.

Hydrangea has diuretic properties and is said to have been much used by the Cherokees and early settlers in calculous complaints.

Oregon Grape.

Berberis Aquifolium Pursi

Pharmacopeial Name--Berberis.

Other Common Names--Rocky Mountain grape, holly-leaved barberry, California barberry, trailing Mahonia.

Habitat and Range--This shrub is native in woods in rich soil among rocks from Colorado to the Pacific Ocean, but is especially abundant in Oregon and northern California.

[Illustration: Oregon Grape (Berberis Aquifolium).]

Description of Plant--Oregon grape is a low-growing shrub, resembling somewhat the familiar Christmas holly of the Eastern states, and, in fact, was first designated as "mountain-holly" by members of the Lewis and Clark expedition on their way through the western country. It belongs to the barberry family (Berberidaceae), and grows about 2 to 6 feet in height, the branches sometimes trailing. The leaves consist of from 5 to 9 leaflets, borne in pairs, with an odd leaflet at the summit. They are from 2 to 3 inches long and about 1 inch wide, evergreen, thick, leathery, oblong or oblong ovate in outline, smooth and shining above, the margins provided with thorny spines or teeth. The numerous small yellow flowers appear in April or May and are borne in erect, clustered heads. The fruit consists of a cluster of blue or bluish purple berries, having a pleasant taste, and each containing from three to nine seeds.

Other Species--While Berberis aquifolium is generally designated as the source of Oregon grape root, other species of Berberis are met with in the market under the name grape root, and their use is

CHAPTER XXXV.

sanctioned by the United States Pharmacopoeia.

The species most commonly collected with Berberis aquifolium is B. nervosa Pursh, which is also found in woods from California northward to Oregon and Washington. This is 9 to 17 inches in height, with a conspicuously jointed stem and 11 to 17 bright-green leaflets.

Another species of Berberis, B. pinnata Lag., attains a height of from a few inches to 5 feet, with from 5 to 9, but sometimes more, leaflets, which are shining above and paler beneath. This resembles aquifolium very closely and is often mistaken for it, but it is said that it has not been used by the medical profession, unless in local practice. The root also is about the same size as that of aquifolium, while the root of nervosa is smaller.

Some works speak of Berberis repens Lindl. as another species often collected with aquifolium, but in the latest botanical manuals no such species is recognized, B. repens being given simply as a synonym for B. aquifolium.

Description of Rootstock--The rootstock and roots of Oregon grape are more or less knotty, in irregular pieces of varying lengths, and about an inch or less in diameter, with brownish bark and hard and tough yellow wood, showing a small pith and narrow rays. Oregon grape root has a very bitter taste and very slight odor.

Collection, Prices and Uses--Oregon grape root is collected in autumn and brings from 10 to 12 cents a pound. The bark should not be removed from the rootstocks, as the Pharmacopoeia directs that such roots be rejected.

This root has long been used in domestic practice thruout the West as a tonic and blood purifier and is now official in the United States Pharmacopoeia.

The berries are used in making preserves and cooling drinks.

END OF GINSENG AND OTHER MEDICINAL PLANTS

CHAPTER XXXV.

End of the Project Gutenberg EBook of Ginseng and Other Medicinal Plants, by A. R. (Arthur Robert) Harding

*** END OF THIS PROJECT GUTENBERG EBOOK GINSENG AND OTHER MEDICINAL PLANTS ***

***** This file should be named 34570.txt or 34570.zip ***** This and all associated files of various formats will be found in:
http://www.gutenberg.org/3/4/5/7/34570/

Produced by Linda M. Everhart, Blairstown, Missouri (This file was produced from images generously made available by The Internet Archive/American Libraries.)

Updated editions will replace the previous one--the old editions will be renamed.

Creating the works from public domain print editions means that no one owns a United States copyright in these works, so the Foundation (and you!) can copy and distribute it in the United States without permission and without paying copyright royalties. Special rules, set forth in the General Terms of Use part of this license, apply to copying and distributing Project Gutenberg-tm electronic works to protect the PROJECT GUTENBERG-tm concept and trademark. Project Gutenberg is a registered trademark, and may not be used if you charge for the eBooks, unless you receive specific permission. If you do not charge anything for copies of this eBook, complying with the rules is very easy. You may use this eBook for nearly any purpose such as creation of derivative works, reports, performances and research. They may be modified and printed and given away--you may do practically ANYTHING with public domain eBooks. Redistribution is subject to the trademark license, especially commercial redistribution.

*** START: FULL LICENSE ***

THE FULL PROJECT GUTENBERG LICENSE PLEASE READ THIS BEFORE YOU DISTRIBUTE OR USE THIS WORK

CHAPTER XXXV.

To protect the Project Gutenberg-tm mission of promoting the free distribution of electronic works, by using or distributing this work (or any other work associated in any way with the phrase "Project Gutenberg"), you agree to comply with all the terms of the Full Project Gutenberg-tm License (available with this file or online at http://gutenberg.org/license).

Section 1. General Terms of Use and Redistributing Project Gutenberg-tm electronic works

1.A. By reading or using any part of this Project Gutenberg-tm electronic work, you indicate that you have read, understand, agree to and accept all the terms of this license and intellectual property (trademark/copyright) agreement. If you do not agree to abide by all the terms of this agreement, you must cease using and return or destroy all copies of Project Gutenberg-tm electronic works in your possession. If you paid a fee for obtaining a copy of or access to a Project Gutenberg-tm electronic work and you do not agree to be bound by the terms of this agreement, you may obtain a refund from the person or entity to whom you paid the fee as set forth in paragraph 1.E.8.

1.B. "Project Gutenberg" is a registered trademark. It may only be used on or associated in any way with an electronic work by people who agree to be bound by the terms of this agreement. There are a few things that you can do with most Project Gutenberg-tm electronic works even without complying with the full terms of this agreement. See paragraph 1.C below. There are a lot of things you can do with Project Gutenberg-tm electronic works if you follow the terms of this agreement and help preserve free future access to Project Gutenberg-tm electronic works. See paragraph 1.E below.

1.C. The Project Gutenberg Literary Archive Foundation ("the Foundation" or PGLAF), owns a compilation copyright in the collection of Project Gutenberg-tm electronic works. Nearly all the individual works in the collection are in the public domain in the United States. If an individual work is in the public domain in the United States and you are located in the United States, we do not claim a right to prevent you from copying, distributing, performing, displaying or creating derivative

CHAPTER XXXV.

works based on the work as long as all references to Project Gutenberg are removed. Of course, we hope that you will support the Project Gutenberg-tm mission of promoting free access to electronic works by freely sharing Project Gutenberg-tm works in compliance with the terms of this agreement for keeping the Project Gutenberg-tm name associated with the work. You can easily comply with the terms of this agreement by keeping this work in the same format with its attached full Project Gutenberg-tm License when you share it without charge with others.

1.D. The copyright laws of the place where you are located also govern what you can do with this work. Copyright laws in most countries are in a constant state of change. If you are outside the United States, check the laws of your country in addition to the terms of this agreement before downloading, copying, displaying, performing, distributing or creating derivative works based on this work or any other Project Gutenberg-tm work. The Foundation makes no representations concerning the copyright status of any work in any country outside the United States.

1.E. Unless you have removed all references to Project Gutenberg:

1.E.1. The following sentence, with active links to, or other immediate access to, the full Project Gutenberg-tm License must appear prominently whenever any copy of a Project Gutenberg-tm work (any work on which the phrase "Project Gutenberg" appears, or with which the phrase "Project Gutenberg" is associated) is accessed, displayed, performed, viewed, copied or distributed:

This eBook is for the use of anyone anywhere at no cost and with almost no restrictions whatsoever. You may copy it, give it away or re-use it under the terms of the Project Gutenberg License included with this eBook or online at www.gutenberg.org

1.E.2. If an individual Project Gutenberg-tm electronic work is derived from the public domain (does not contain a notice indicating that it is posted with permission of the copyright holder), the work can be copied and distributed to anyone in the United States without paying any fees or charges. If you are redistributing or providing access to a work with the phrase "Project Gutenberg" associated with or appearing on the work, you must comply either with the requirements of paragraphs 1.E.1 through 1.E.7 or obtain permission for the use of the work and the Project Gutenberg-tm trademark as set forth in paragraphs 1.E.8 or 1.E.9.

1.E.3. If an individual Project Gutenberg-tm electronic work is posted with the permission of the copyright holder, your use and distribution must comply with both paragraphs 1.E.1 through 1.E.7 and any additional terms imposed by the copyright holder. Additional terms will be linked to the Project Gutenberg-tm License for all works posted with the permission of the copyright holder found at the beginning of this work.

1.E.4. Do not unlink or detach or remove the full Project Gutenberg-tm License terms from this work, or any files containing a part of this work or any other work associated with Project Gutenberg-tm.

1.E.5. Do not copy, display, perform, distribute or redistribute this electronic work, or any part of this electronic work, without prominently displaying the sentence set forth in paragraph 1.E.1 with active links or immediate access to the full terms of the Project Gutenberg-tm License.

1.E.6. You may convert to and distribute this work in any binary, compressed, marked up, nonproprietary or proprietary form, including any word processing or hypertext form. However, if you provide access to or distribute copies of a Project Gutenberg-tm work in a format other than "Plain Vanilla ASCII" or other format used in the official version posted on the official Project Gutenberg-tm web site (www.gutenberg.org), you must, at no additional cost, fee or expense to the user, provide a copy, a means of exporting a copy, or a means of obtaining a copy upon request, of the work in its original "Plain Vanilla ASCII" or other form. Any alternate format must include the full Project Gutenberg-tm License as specified in paragraph 1.E.1.

1.E.7. Do not charge a fee for access to, viewing, displaying, performing, copying or distributing any Project Gutenberg-tm works unless you comply with paragraph 1.E.8 or 1.E.9.

CHAPTER XXXV.

1.E.8. You may charge a reasonable fee for copies of or providing access to or distributing Project Gutenberg-tm electronic works provided that

- You pay a royalty fee of 20% of the gross profits you derive from the use of Project Gutenberg-tm works calculated using the method you already use to calculate your applicable taxes. The fee is owed to the owner of the Project Gutenberg-tm trademark, but he has agreed to donate royalties under this paragraph to the Project Gutenberg Literary Archive Foundation. Royalty payments must be paid within 60 days following each date on which you prepare (or are legally required to prepare) your periodic tax returns. Royalty payments should be clearly marked as such and sent to the Project Gutenberg Literary Archive Foundation at the address specified in Section 4, "Information about donations to the Project Gutenberg Literary Archive Foundation."

- You provide a full refund of any money paid by a user who notifies you in writing (or by e-mail) within 30 days of receipt that s/he does not agree to the terms of the full Project Gutenberg-tm License. You must require such a user to return or destroy all copies of the works possessed in a physical medium and discontinue all use of and all access to other copies of Project Gutenberg-tm works.

- You provide, in accordance with paragraph 1.F.3, a full refund of any money paid for a work or a replacement copy, if a defect in the electronic work is discovered and reported to you within 90 days of receipt of the work.

- You comply with all other terms of this agreement for free distribution of Project Gutenberg-tm works.

1.E.9. If you wish to charge a fee or distribute a Project Gutenberg-tm electronic work or group of works on different terms than are set forth in this agreement, you must obtain permission in writing from both the Project Gutenberg Literary Archive Foundation and Michael Hart, the owner of the Project Gutenberg-tm trademark. Contact the Foundation as set forth in Section 3 below.

1.F.

1.F.1. Project Gutenberg volunteers and employees expend considerable effort to identify, do copyright research on, transcribe and proofread public domain works in creating the Project Gutenberg-tm collection. Despite these efforts, Project Gutenberg-tm electronic works, and the medium on which they may be stored, may contain "Defects," such as, but not limited to, incomplete, inaccurate or corrupt data, transcription errors, a copyright or other intellectual property infringement, a defective or damaged disk or other medium, a computer virus, or computer codes that damage or cannot be read by your equipment.

1.F.2. LIMITED WARRANTY, DISCLAIMER OF DAMAGES - Except for the "Right of Replacement or Refund" described in paragraph 1.F.3, the Project Gutenberg Literary Archive Foundation, the owner of the Project Gutenberg-tm trademark, and any other party distributing a Project Gutenberg-tm electronic work under this agreement, disclaim all liability to you for damages, costs and expenses, including legal fees. YOU AGREE THAT YOU HAVE NO REMEDIES FOR NEGLIGENCE, STRICT LIABILITY, BREACH OF WARRANTY OR BREACH OF CONTRACT EXCEPT THOSE PROVIDED IN PARAGRAPH 1.F.3. YOU AGREE THAT THE FOUNDATION, THE TRADEMARK OWNER, AND ANY DISTRIBUTOR UNDER THIS AGREEMENT WILL NOT BE LIABLE TO YOU FOR ACTUAL, DIRECT, INDIRECT, CONSEQUENTIAL, PUNITIVE OR INCIDENTAL DAMAGES EVEN IF YOU GIVE NOTICE OF THE POSSIBILITY OF SUCH DAMAGE.

1.F.3. LIMITED RIGHT OF REPLACEMENT OR REFUND - If you discover a defect in this electronic work within 90 days of receiving it, you can receive a refund of the money (if any) you paid for it by sending a written explanation to the person you received the work from. If you received the work on a physical medium, you must return the medium with your written explanation. The person or entity that provided you with the defective work may elect to provide a replacement copy in lieu of a refund. If you received the work electronically, the person or entity providing it to you may choose to give you a second opportunity to receive the work electronically in lieu of a refund. If the second copy is also defective, you may demand a refund in writing without further opportunities to fix the problem.

CHAPTER XXXV. 241

1.F.4. Except for the limited right of replacement or refund set forth in paragraph 1.F.3, this work is provided to you 'AS-IS' WITH NO OTHER WARRANTIES OF ANY KIND, EXPRESS OR IMPLIED, INCLUDING BUT NOT LIMITED TO WARRANTIES OF MERCHANTIBILITY OR FITNESS FOR ANY PURPOSE.

1.F.5. Some states do not allow disclaimers of certain implied warranties or the exclusion or limitation of certain types of damages. If any disclaimer or limitation set forth in this agreement violates the law of the state applicable to this agreement, the agreement shall be interpreted to make the maximum disclaimer or limitation permitted by the applicable state law. The invalidity or unenforceability of any provision of this agreement shall not void the remaining provisions.

1.F.6. **INDEMNITY**

- You agree to indemnify and hold the Foundation, the trademark owner, any agent or employee of the Foundation, anyone providing copies of Project Gutenberg-tm electronic works in accordance with this agreement, and any volunteers associated with the production, promotion and distribution of Project Gutenberg-tm electronic works, harmless from all liability, costs and expenses, including legal fees, that arise directly or indirectly from any of the following which you do or cause to occur: (a) distribution of this or any Project Gutenberg-tm work, (b) alteration, modification, or additions or deletions to any Project Gutenberg-tm work, and (c) any Defect you cause.

Section 2. Information about the Mission of Project Gutenberg-tm

Project Gutenberg-tm is synonymous with the free distribution of electronic works in formats readable by the widest variety of computers including obsolete, old, middle-aged and new computers. It exists because of the efforts of hundreds of volunteers and donations from people in all walks of life.

Volunteers and financial support to provide volunteers with the assistance they need, are critical to reaching Project Gutenberg-tm's goals and ensuring that the Project Gutenberg-tm collection will remain freely available for generations to come. In 2001, the Project

CHAPTER XXXV.

Gutenberg Literary Archive Foundation was created to provide a secure and permanent future for Project Gutenberg-tm and future generations. To learn more about the Project Gutenberg Literary Archive Foundation and how your efforts and donations can help, see Sections 3 and 4 and the Foundation web page at http://www.pglaf.org.

Section 3. Information about the Project Gutenberg Literary Archive Foundation

The Project Gutenberg Literary Archive Foundation is a non profit 501(c)(3) educational corporation organized under the laws of the state of Mississippi and granted tax exempt status by the Internal Revenue Service. The Foundation's EIN or federal tax identification number is 64-6221541. Its 501(c)(3) letter is posted at http://pglaf.org/fundraising. Contributions to the Project Gutenberg Literary Archive Foundation are tax deductible to the full extent permitted by U.S. federal laws and your state's laws.

The Foundation's principal office is located at 4557 Melan Dr. S. Fairbanks, AK, 99712., but its volunteers and employees are scattered throughout numerous locations. Its business office is located at 809 North 1500 West, Salt Lake City, UT 84116, (801) 596-1887, email business@pglaf.org. Email contact links and up to date contact information can be found at the Foundation's web site and official page at http://pglaf.org

For additional contact information: Dr. Gregory B. Newby Chief Executive and Director gbnewby@pglaf.org

Section 4. Information about Donations to the Project Gutenberg Literary Archive Foundation

Project Gutenberg-tm depends upon and cannot survive without wide spread public support and donations to carry out its mission of increasing the number of public domain and licensed works that can be freely distributed in machine readable form accessible by the widest array of equipment including outdated equipment. Many small donations ($1 to $5,000) are particularly important to maintaining tax

CHAPTER XXXV.

exempt status with the IRS.

The Foundation is committed to complying with the laws regulating charities and charitable donations in all 50 states of the United States. Compliance requirements are not uniform and it takes a considerable effort, much paperwork and many fees to meet and keep up with these requirements. We do not solicit donations in locations where we have not received written confirmation of compliance. To SEND DONATIONS or determine the status of compliance for any particular state visit http://pglaf.org

While we cannot and do not solicit contributions from states where we have not met the solicitation requirements, we know of no prohibition against accepting unsolicited donations from donors in such states who approach us with offers to donate.

International donations are gratefully accepted, but we cannot make any statements concerning tax treatment of donations received from outside the United States. U.S. laws alone swamp our small staff.

Please check the Project Gutenberg Web pages for current donation methods and addresses. Donations are accepted in a number of other ways including checks, online payments and credit card donations. To donate, please visit: http://pglaf.org/donate

Section 5. General Information About Project Gutenberg-tm electronic works.

Professor Michael S. Hart is the originator of the Project Gutenberg-tm concept of a library of electronic works that could be freely shared with anyone. For thirty years, he produced and distributed Project Gutenberg-tm eBooks with only a loose network of volunteer support.

Project Gutenberg-tm eBooks are often created from several printed editions, all of which are confirmed as Public Domain in the U.S. unless a copyright notice is included. Thus, we do not necessarily keep eBooks in compliance with any particular paper edition.

CHAPTER XXXV.

Most people start at our Web site which has the main PG search facility:

http://www.gutenberg.org

This Web site includes information about Project Gutenberg-tm, including how to make donations to the Project Gutenberg Literary Archive Foundation, how to help produce our new eBooks, and how to subscribe to our email newsletter to hear about new eBooks.

Ginseng and Other Medicinal Plants, by

A free ebook from http://manybooks.net/

www.ingramcontent.com/pod-product-compliance
Lightning Source LLC
Chambersburg PA
CBHW050052230526
45470CB00004B/1499